METAL CASTING: A SAND CASTING MANUAL FOR THE SMALL FOUNDRY VOL. II

STEPHEN D. CHASTAIN

B.SC. MECHANICAL ENGINEERING AND MATERIALS SCIENCE
UNIVERSITY OF CENTRAL FLORIDA

Metal Casting: A Sand Casting Manual
for the Small Foundry Vol. II
By Stephen D. Chastain

Copyright© 2004 By Stephen D. Chastain

Jacksonville, FL All Rights Reserved

Printed in USA

ISBN 0-9702203-3-2

The Small Foundry Series by Stephen Chastain As of 2005:

Volume I.   Iron Melting Cupola Furnaces for the Small Foundry

Volume II. Build an Oil-Fired Tilting Furnace

Volume III. Metal Casting: A Sand Casting Manual Vol. I

Volume IV. Metal Casting: A Sand Casting Manual Vol. II

Volume V. Making Pistons for Experimental and Restoration Engines

Volume VI. Sand Machinery

Other Books by Stephen Chastain:

Generators and Inverters: Building Small Combined Heat and Power Plants

Web Site: **http://StephenChastain.Com** Stevechastain@hotmail.com

Steve Chastain

2925 Mandarin Meadows Dr.

Jacksonville, FL 32223

TABLE OF CONTENTS VOLUME 2

CONTINUED FROM VOLUME I

I. SOLIDIFICATION OF METALS   6

    Freezing of Pure Metals   8

    Freezing of Alloys   9

    Phase Diagrams   16

II. ALUMINUM ALLOYS: 21

    The Effect of Alloying Agents   21

    Grain Refiners   23

    Solution Heat Treatment & Precipitation

    Hardening   25

    The Composition of Aluminum Alloys   27

    Melt Reactions   32

    Aluminum Melting Practice   35

III. COPPER ALLOYS, BRASS & BRONZE   42

    Copper-Based Casting Alloys   45

    Silicon Bronze   45

    Brasses   46

    Manganese Bronze   47

    Tin and Leaded Bronze   50

    Aluminum Bronze   54

    Solidification Range of Copper Alloys   58

    Brass and Bronze Foundry Practice   59

IV. METALLURGY OF IRON 65

    Carbon 68

    Silicon 69

    Carbon Equivalent 70

    Alloying Elements 72

    Ductile Iron 74

    Alkali Fluxes 75

V. GATING SYSTEMS 77

    Parts of a Gating System 77

    Filters 87

    Design of Gating Systems 88

VI. RISERS AND FEEDING OF CASTINGS 99

    Directional Solidification 104

    Chills 105

    Heat Loss From Risers 108

    Making Insulated Riser Sleeves 111

VII. PATTERN MAKING 117

    Shrinkage 117

    Machining Allowance 118

    Fillets 119

    Wax Extruder 120

    Follow Boards 121

    Core Prints 124

    Stop Offs 125

Rubber molds   129

Building a Vacuum Chamber   137

Making a Match Plate   140

VIII. FOUNDRY PROJECTS   148

  Make a Sturdy Flask Lock   148

  Make an Knee Operated Air Valve   152

  Cylinder Head Casting   160

  Piston Casting   169

IX. AUTOMOTIVE CASTINGS   172

  Casting Piston Rings   172

  Production of the Ford Flathead V-8   178

X. MISCLANENIOUS   185

  Molding Deep Fins   185

  Lifting Force on Cope   186

XI. CONCLUSION   187

BIBLIOGRAPHY   188

SUPPLIERS   189

COMPOSITION OF ALUMINUM ALLOYS   190

INDEX   191

I. SOLIDIFICATION OF METALS:

The change from molten metal to a cool solid casting takes place in three main steps. First the cooling of the metal from the pouring temperature to the solidification temperature. The difference between the pouring temperature and the solidification temperature is called the superheat. The second step is removal of the heat of fusion through the solidification range. The third step is cooling of the solid casting to room temperature.

Before the description of metal solidification, it is important to define a few terms.

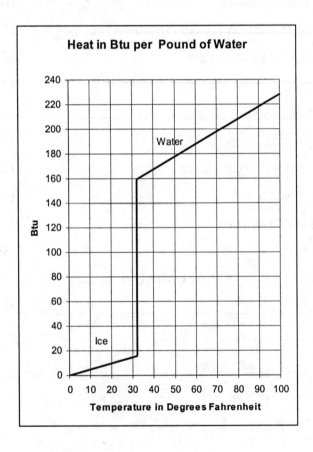

When water freezes, heat is removed yet the temperature of the water remains at 32°F until all of the water turns to ice. The amount of heat that has to be removed to freeze a substance *after* it reaches the freezing temperature is called the **heat of fusion**. In water and pure metals, there is no change in temperature until the change from liquid to solid is complete. For water, the heat of fusion is 144 Btu per pound. The heat of fusion of various metals is found in the table of properties.

The **state** of matter or **phase** refers to the condition it is in such as solid, liquid or gas.

Precipitate refers to the spontaneous generation of one phase from a mixture of another phase. Rain precipitates from a mixture of air and water vapor. A solid metal or a component of a solution may precipitate from a liquid solution of metals or alloying agents.

Eutectic (you-tek-tik) refers to a combination of metals or alloying agents that freeze at a constant temperature like water. A eutectic combination has the lowest freezing point of any possible combination of the alloying elements and the freezing point is lower than any of the individual constituents of the alloy. A eutectic precipitates two or more phases simultaneously. Eutectic grain size is much finer than that of a pure metal.

Metals solidify by one of three methods:

1. At a constant temperature (like water). Pure metals and eutectic alloys solidify at constant temperature.
2. Over a temperature range. Solid solutions solidify over a temperature range
3. By a combination of solidification over a temperature range and then by constant temperature freezing.

7

Freezing of a Pure Metal:

When a pure metal freezes in a mold, a thin skin of solid metal forms around the liquid. More liquid begins to freeze on top of the skin and it becomes thicker, growing towards the center of the casting. The layer between the liquid and the solid is relatively smooth because the metal is freezing at a constant temperature. As the metal solidifies, heat is conducted through the mold wall and the heat of fusion is released into the remaining liquid, keeping the temperature constant. The release of the heat of fusion slows the nucleation of new grains. As more heat is removed through the mold wall, atoms attach themselves to the existing nuclei causing the grains to grow while forming new nuclei. The gradual inward growth of grains from the surface of a casting does not restrict the flow of feed metal in the liquid interior of the casting.

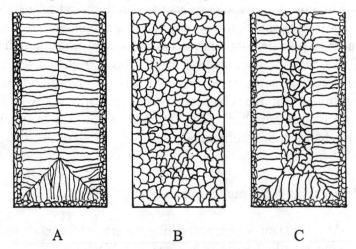

A   B   C

The part of the casting that is closest to the wall solidifies as fine equi-axed grains. Equi-axed grains are the same size in all directions as opposed to long and thin. Grain growth occurs opposite the direction the heat flow. Only those grains pointed towards the center of the casting will grow with the others being pinched off as seen in drawing A. This creates a region of columnar grains next to

8

the outer layer of fine grains. In pure metals, these columnar grains extend to the center of the casting, while in alloys they may be interrupted by equi-axed grains when the alloy reaches a eutectic composition as seen in drawing C.

Drawing B represents equi-axed grains developed in long freezing range alloys, or an alloy that has been modified.

Freezing of Alloys:

Pure metals freeze at a constant temperature. Alloys, in most cases, do not. For a given alloy there is a particular temperature called the liquidus, above which the metal is completely liquid; there is also a lower temperature where the metal is completely solid called the solidus. Between these two temperatures is a mushy or pasty phase where both the solid and liquid coexist. This range of temperatures is called the solidification range. Freezing starts at the liquidus temperature and is completed at the solidus temperature.

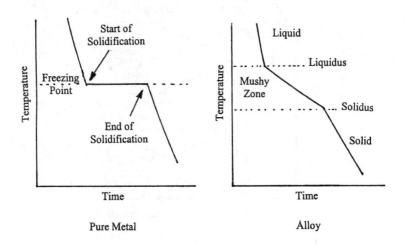

Pure Metal   Alloy

9

Alloys can be divided into groups that:

1. Freeze as solid solutions.
2. Freeze by precipitating an essentially pure component.
3. Freeze at a constant temperature. This is called *eutectic* freezing where the alloy precipitates 2 or more phases simultaneously.
4. Starting as #1 or #2, and ending as #3.

Types 1 and 2:

As solidification progresses, dendrites grow by extending their spines into the liquid metal and sending out branches similar to a fir-tree. In the alloy, the metal with the highest melting temperature first freezes on these spines thereby changing the composition of the remaining liquid. The last metal to freeze around the dendrite is usually very different in composition than the first metal to freeze.

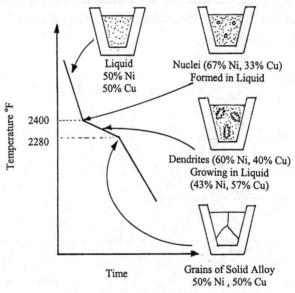

Solid Solution Freezing
50% Nickel 50% CopperAlloy

10

This creates layers like the skin of an onion around the dendrites. As the dendrites grow, they block the channels through which liquid metal flows. If the solidification range is large, the distance through which the feed metal must flow through a mushy zone of dendrites is quite long. This resistance to flow is a source of micro porosity.

When a casting cools, the shrinkage occurs at the last place to freeze. Shrinkage also occurs at the last metal to freeze around the dendrites as seen above. This results in dispersed micro porosity as commonly seen in tin bronze. If gasses such as hydrogen are dissolved in the liquid metal, they will accumulate in these spaces between the dendrites, increasing the micro porosity. If the level of gas is high enough, it will create back pressure preventing the feed metal from flowing.

While microshrinkage may be a problem in pressure tight castings, it is not necessarily a problem in others. Some melts may be slightly gassed to ensure dispersed shrinkage. In bearing applications, microshrinkage may act as oil retaining cavities.

Eutectic freezing occurs at a constant temperature as with pure metals. Only one specific composition of an alloy will freeze as a eutectic. The eutectic composition has the lowest freezing point of all the possible combinations of alloying materials.

Although eutectic freezing occurs at a constant temperature like pure metals, there are some differences. The grain size of a eutectic alloy is much finer than that of a pure metal, and freezing starts at a lower temperature than that of either component in the alloy. The shape of the precipitate may be lamellar (form in layers), rodlike or globlualar.

Eutectic alloys solidify either from the surface to the center of the casting in a wall like fashion as seen in pure

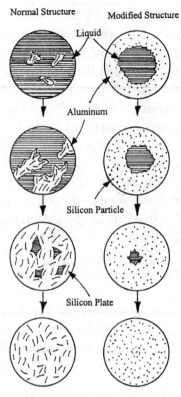

Normal Structure / Modified Structure

Liquid

Aluminum

Silicon Particle

Silicon Plate

Effects of Sodium Modification
of Aluminum-Silicon Alloy

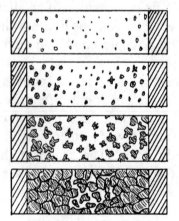

Long Freezing Range Alloy

metals, or at random locations throughout the liquid creating a "mushy" condition. Aluminum silicon alloys solidify randomly; however the addition of sodium modifies the alloy causing the alloy to solidify from the surface to the center.

Aluminum castings freeze by three different methods. In pure aluminum, shrinkage occurs as a deep pipe or at the centerline of the casting. Solidification is as was described above, for pure metals.

Solidification of alloy #295, 94% aluminum, 5% copper, 1% silicon begins at the wall but progresses quickly to the center of the casting. Fine grains form randomly in the center of the casting and freezing continues in a mushy state. The center of the casting may be as much as 85% solid before a completely solid skin forms on the surface. As a network of solid grains form, feed metal is unable to flow through the constricted passages and

12

microshrinkage occurs around the dendrites. The riser height drops and distributed microshrinkage forms throughout the riser and casting.

Chills are used to force the metal to freeze quickly from one end before the network of grains forms, constricting the flow of feed metal. Chills also increase the mechanical properties by reducing the segregation of gas and impurities at the grain boundaries.

Alloy #332, silicon 9.5%, copper 3% solidifies with some gross shrinkage and some distributed microshrinkage.

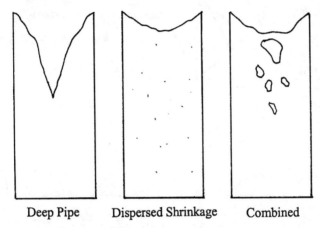

Deep Pipe Dispersed Shrinkage Combined

Many metals solidify by a combination of solidification over a temperature range and then by constant temperature freezing. Solidification starts by growth of dendrites of nearly pure metal followed by the eutetic solidification of the remaining liquid. Examples of aluminum copper and aluminum silicon alloys are given. Gray iron also solidifies by this method.

Gray Iron: If the carbon equivalent is less than 4.3*, the iron is hypoeutectic and starts solidification by growing pure iron dendrites. When enough iron has solidified from the solution for it to reach the eutectic composition, both graphite and iron precipitate simultaneously.

*See the chapter regarding metallurgy of iron.

If the carbon equivalent is greater than 4.3, the iron is hypereutectic and solidification begins by the precipitation of graphite. When the eutectic composition is reached, eutectic solidification occurs.

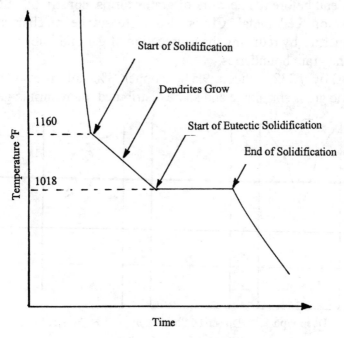

Aluminum –10% Copper Alloy

Combination of Solidification over a Temperature Range Followed by Constant Temperature Freezing

Copper alloys may have a short freezing range and solidify similar to pure metals. These include silicon bronze, aluminum bronze and manganese bronze. Shrinkage occurs along the centerline, at hotspots, and as riser piping.

Long freezing range alloys, like some aluminum alloys, solidify in a mushy state. Dendrites block the flow of feed metal, therefore increasing the riser size does not

14

necessarily improve the dispersed shrinkage. Chilling to set up directional solidification is more effective in eliminating shrinkage. Long freezing range alloys include red brass, some yellow brasses, gunmetal and leaded gunmetal.

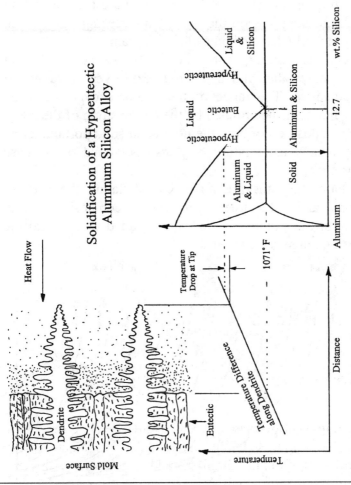

Solidification of an Aluminum Silicon Alloy by a Combination of Solidification over a Temperature Range Followed by Constant Temperature Solidification

15

PHASE DIAGRAMS:

You can go through your whole life pouring castings and not see a phase diagram. They are not essential for making castings; however they are common in literature regarding metals and metallurgy. They are a good way to communicate ideas on the topic. The freezing range of an alloy is easily detected on a phase diagram.

As seen earlier, pure metals freeze at a constant temperature and alloys may freeze over a range of temperatures. The range of temperatures over which an alloy freezes is determined by the composition of the alloy. The composition of both the liquid and solid portions of an alloy at any temperature may also be found on a phase diagram.

Phase diagrams are made by tabulating the start of freezing and end of freezing for various compositions of an alloy. In the example below, tin is added to lead and the freezing range is recorded.

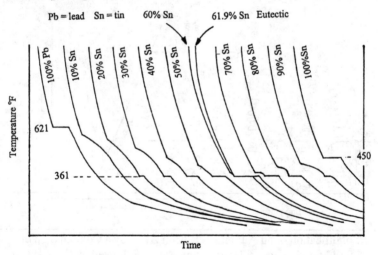

Cooling Curves for Lead and Tin

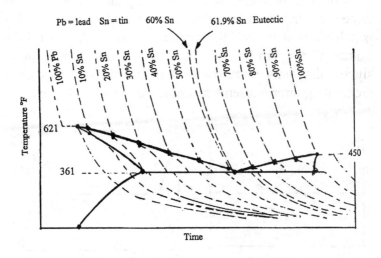

The start and end of freezing of each composition are connected with a line and the time axis is converted to weight percent composition as seen below.

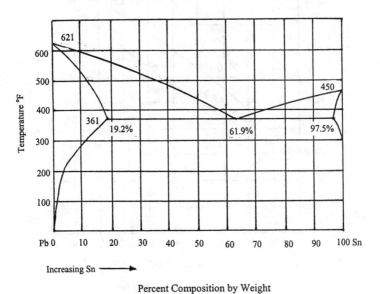

Percent Composition by Weight

Lead –Tin Phase Diagram

17

Although there are several types of phase diagrams, only three are introduced here. If neither constituent of the alloy is soluble in the other, the system is seen in the "No Solid Solution" diagram below. The lead-tin and the aluminum-silicon diagrams represent partial solubility. The copper nickel diagram represents complete solubility.

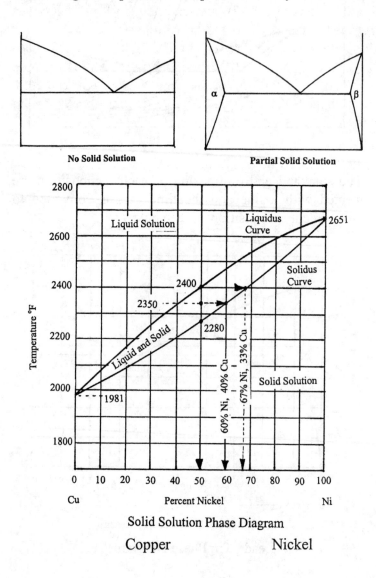

No Solid Solution   Partial Solid Solution

Solid Solution Phase Diagram

Copper   Nickel

18

The composition of both the solid and liquid phases at any temperature may be determined from the phase diagram. Using a mixture of 50% copper and 50% nickel at a temperature of 2400 °F, the solid phase consists of 67% nickel and 33% copper. At 2350 °F, the solid phase consists of 60% nickel and 40% copper and at 2280 °F the solid phase consists of 50% copper and 50% nickel. This phase diagram corresponds to the time-temperature diagram for solid solution freezing as seen in the solidification chapter.

An aluminum, 10% copper alloy solidifies over a temperature range followed by constant temperature solidification. The time temperature diagram is also found in the solidification chapter.

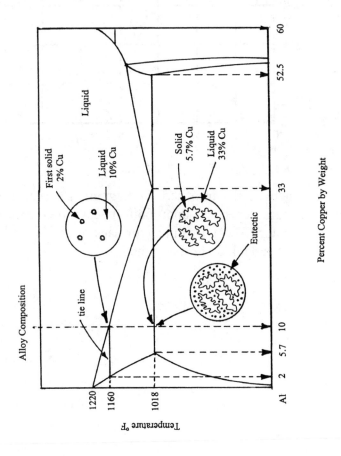

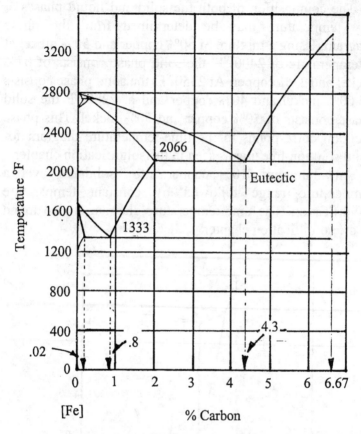

Seen here is the iron carbon diagram. Because iron's properties are dependent upon both its carbon content and thermal history (heat treatment or lack of it), it is best discussed in a book on heat treatment.

II. ALUMINUM ALLOYS:

There are two main types of aluminum alloys, wrought alloys and casting alloys. Wrought aluminum is used for extruded shapes such as aluminum cans, window frames, tubing, and angles to name a few. Their properties are tailored for die forging, not casting work. While I have melted and cast wrought alloys, you may or may not get a simple workable casting. If you want metal to flow properly into more intricate molds and you want certain engineering properties of the finished casting, you are much better off to use casting alloys. Casting alloys are of two types; alloys that are used in the "as cast" condition and heat treatable alloys.

Pure aluminum is a poor or less desirable casting material. It is alloyed with one or more of the following: copper, silicon, magnesium, zinc, chromium, nickel, titanium, or tin. Iron may be present in some die casting alloys, but it is generally considered to be an impurity in other alloys. Alloys may be tailored for "as cast" strength, hot strength, low thermal expansion, pressure tightness, ductility, or fluidity. They may also be formulated for heat-treatment.

Copper was the earliest used alloying material. Copper percentages in aluminum casting alloys are usually between 2 and 12 percent. The hardness and strength increase as the copper is increased; however the ductility decreases with increasing copper. Ductile or tough alloys use between 2 to 5 percent copper. Increasing amounts of copper are used for higher strength and hardness. Early casting alloys used 8 percent copper and were used in the as cast condition. Later the percentage was reduced to 4 per cent to give a heat treatable alloy that was also ductile. Copper also increases the hot strength of the alloy.

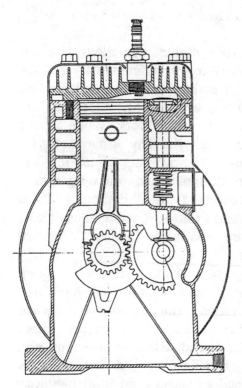

Silicon is used up to about 18 percent in casting alloys. It greatly increases the fluidity. This is due to the fact that silicon's heat of fusion is 775 Btu per pound. As it solidifies, it releases a great amount of heat to the liquid aluminum keeping it fluid.

The alloy used for the die casting on the left requires a high silicon content to properly fill the thin sections.

Addition of silicon also increases the strength up to the eutectic point (12.7% silicon) Silicon reduces the thermal expansion of the finished casting and it also reduces solidification shrinkage. Corrosion resistance is improved and weldability is good. Silicon also combines with magnesium to forms Mg_2Si, which makes a heat treatable alloy.

Slow solidification produces a coarse grain size in aluminum silicon alloys. Sand casting therefore produces a coarse grain size and the alloys are often modified by using a "grain refiner" such as titanium boron or sodium. Aluminum silicon alloys are best when they are cast in metal molds or modified as mentioned above. Coarse grain sizes make a brittle material. Rapid cooling, as seen in permanent mold casting, results in a fine grain and improved ductility. Iron causes coarse grain size and brittle castings.

Magnesium produces effects similar to those of copper. It is added from 4 to 10 percent in the binary (two metal) alloys. Aluminum magnesium alloys are more difficult to cast than aluminum silicon alloys. They have a tendency to oxidize in the molten state. Steam generated by sand casting reacts with the magnesium to form MgO and hydrogen. This gives a rough blackened surface to the castings. This reaction is controlled by adding about 1.5% boric acid to the sand.

Magnesium and Silicon combine to form Mg_2Si. These alloys are age hardened by precipitation of Mg_2Si. Small amounts of magnesium may more than double the yield strength of an alloy with an equivalent amount of silicon. Alloy number 356 is a popular sand casting alloy with 7% silicon and .35% Magnesium.

Zinc is added to get the maximum hardness in the as cast condition.

GRAIN REFINERS:

The best strength and ductility are found in fine grain castings. Rapid cooling from temperatures above 750° F or a grain refining flux may be added to produce fine grain. Grain refiners are based on **titanium-boron**. Grain refiners may be added in the form of master alloy or salts (flux). Flux may be added at the beginning of a melt, giving it adequate time to dry. Alloys with a large amount of magnesium do not refine as well as those with lower amounts of magnesium.

Sodium refines the silicon in aluminum alloys that contain from 6 to 13 per cent silicon. It increases the mechanical properties, strength and ductility. Without sodium, silicon forms as coarse plate-like hard crystals in a soft aluminum matrix (upper photo on the next page). When bending this type of structure, the alloy starts to fracture at these crystals. An unmodified alloy may bend less than 20 degrees before breaking. Silicon forms as many small

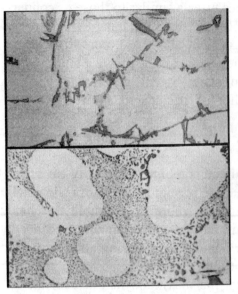

particles with the addition of a small amount of sodium. A sodium-modified alloy (lower photo) may bend 70 to 90 degrees before breaking. Sodium is added at the rate of .02 weight percent as packaged metallic sodium or as sodium salts (flux). Because fluxes absorb water, they are less desirable than straight metallic sodium. Metallic sodium is very reactive and must be handled carefully. It reacts violently with water to form sodium hydroxide and hydrogen. The reaction produces a great amount of heat that causes the released hydrogen to burst into flame.

Phosphorus: Alloys that contain more than 12.7 percent silicon are called "hypereutectic." During solidification, the silicon in these alloys forms large block-like hard particles. Phosphorous additions form many small particles of aluminum phosphide that is finely dispersed in the molten aluminum. The silicon forms on these small particles thus refining the melt. Phosphorus is added as copper phosphorus shot at the rate of .065 percent for sand castings or .035 percent for permanent mold castings. Hypereutectic alloys usually require a two step treatment, titanium-boron to refine the aluminum grain and phosphorus to refine the silicon. Sodium should never be used with a phosphorus treatment because they combine with each other becoming useless as refiner.

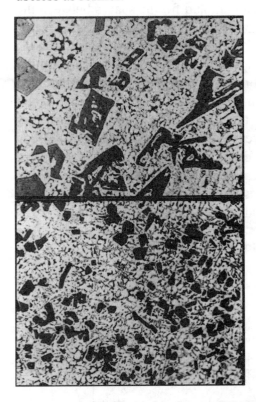

Left: unmodified-top photo, phosphorous modified lower photo.

Solution heat-treatment and Precipitation Hardening:

Copper dissolves in aluminum at melting temperatures, however, it separates out of solution upon slow cooling. To get the maximum strength from a sandcast aluminum copper alloy, it is heat treated. Heat treatment is a two step process where first the casting is heated to 950° to dissolve the copper back into the aluminum. This may take some time so "time at temperature" is an important factor. After the copper has dissolved, the casting is quenched in water to freeze the solution of copper in aluminum. Similar effects may be achieved by chill casting as seen in permanent molds.

Precipitation or Age hardening is the second step in the hardening process. At room temperature, the copper will separate from the aluminum and fill the spaces between the crystals producing a harder hardness. Aging, at an elevated temperature between 300 and 400°F, speeds up the process. Again, time at temperature is critical for maximum hardness. If the casting is heated too long, the hardness exceeds the maximum and the casting starts to get softer. If heating is continued, the casting will return to its soft "as cast" state. The process is similar for magnesium–silicon alloys, however the castings should be slowly quenched from the solution treatment temperature by quenching in boiling water. Magnesium alloys are susceptible to stress-corrosion cracking and require a less drastic quench.

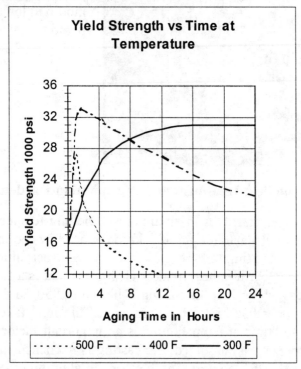

Time at Temperature- Precipitation Hardening

THE COMPOSITON OF ALUMINUM ALLOYS:

The composition and properties of several casting alloys are summarized below. Because aluminum cans are so common, wrought alloy 3004 is mentioned.

Alloy 3004 is used for aluminum cans. It has a tensile strength of 26,000 psi in the annealed condition. It contains 1.2% Manganese, 1% Magnesium and 97.8% Aluminum.

Because Briggs and Stratton engines are common and likely to find their way into a small foundry, the composition of their alloys are given below.

The composition of the alloy used for all aluminum die castings, other than pistons is:

Composition %, Aluminum is the remaining percentage.

Si	Cu	Fe	Mg (max)	Mn (max)	Zn (max)	Sn (max)
11-13	2-3	0. 5-0.7	0.1	0. 5	1	0.15

The alloy used for pistons has the composition below:

Si	Cu	Fe	Mg (max)	Mn (max)	Zn (max)	Sn (max)
5-7.5	2-3	0. 5-0.7	0.1	0. 5	1	0.15

The Aluminum Association alloy numbering system is summarized below. The composition of common alloys are found in the chart at the end of this section.

Series	Type
1XX.X	99% or more aluminum
2XX.X	Al + Cu
3XX.X	Al + (Si-Mg), (Si-Cu) or (Si-Mg-Cu)
4XX.X	Al + Si
5XX.X	Al + Mg
7XX.X	Al + Zn
8XX.X	Al + Sn

The 6XX.X and 9XX.X numbers are not in use at this time.

Some alloys include a letter like A356.0 as opposed to 356.0. Alloy A356.0 has lower impurities that give it higher strength and ductility.

Alloy 208.0 has very good casting characteristics. Fluidity is good and pressure tightness is very good. It is used for manifolds and other parts that must be pressure tight. It is used in the as cast condition. Weldability is good. Finish is very good and it polishes well.

Alloys 242.0 & A242.0 are used where high strength and hardness at high temperatures are important. Applications include air-cooled cylinder heads on aircraft engines, motorcycle pistons, diesel and heavy duty pistons. Machinability is very good and fluidity is good. The alloy is fair regarding hot cracking and solidification shrinkage. Finish and polish are very good.

Alloy 295.0 & 296.0. Number 295.0 was very popular until the mid 1930's. Alloys 355.0 and 356 have better casting characteristics and have replaced much of the work done with 295 and 296. Number 296.0 has higher silicon content than 295.0 that increases fluidity and reduces shrinkage. Applications include gear housings, aircraft and railway fittings, compressor connecting rods. Machinability and weldability are good. They finish and polish well.

Alloys 319.0, A319.0, B319.0 & 320.0 are very good casting alloys. They are weldable and the machinability is good. Solidification shrinkage is low and pressure tightness is very good. Applications include Gasoline and diesel crankcases, water-cooled cylinder heads, and oil pans. Finish is fair.

Alloys 332 & 336 are used when good strength at high temperatures, wear resistance and low expansion at elevated temperatures are needed. Pressure tightness is good. Because of the high silicon content, machinability is fair. Carbide tools hold up better than high-speed steel because of abrasiveness of the silicon. Small amounts of nickel in these alloys increase the hardness at elevated temperatures. Typical applications include automotive and heavy-duty diesel pistons. 332 was known as F132 piston alloy. Finish is poor.

Alloys 355, 356 are general-purpose casting alloys. They have high strength, excellent pressure tightness and castability. Machinability is good. Alloy 356 is more ductile than 355. Typical applications include transmission cases, engine blocks, compressor pistons, blower housings and car wheels. Finish is good.

Alloy 360 is used for thin walled castings. Due to the high silicon content, machinability is fair but the finish is good.

Alloys 383, 384 are used for pistons and components in severe service conditions. They are used for casings with thin walls and large surface areas such as air-cooled cylinder heads. Fluidity is excellent and machinability using carbide tools is good. Finish is good.

Alloy 390 is a hypereutectic alloy that is used in sand casting of automotive cylinder blocks. The silicon crystals give a wear resistant surface that eliminates the need for cast iron cylinder liners. Thin sections may be easily cast and pressure tightness is good. Adding copper phosphorous modifies the grain. Machinability is good when using cast iron grade carbide tools. Cutting fluid is needed. Finish is good.

Alloy 443 is used for sand and die-castings where above average ductility is needed. It has casting characteristics similar to the 356 alloys. Typical applications include carburetor bodies, cooking utensils and marine fittings.

500 Series alloys, some known as Almag-35 have poor casting characteristics. They have good corrosion resistance and excellent machinability. They are often used in cooking utensils and marine fittings and computing devices where dimensional stability is important.

Alloys 771 and 772 develop high strength in the as cast condition without heat treatment. Castability is good and machinability is excellent. These alloys are dimensionally stable, have good toughness and are able to withstand heavy shock. They polish well.

Alloys 850, 851 and 852 are bearing alloys. They are often used for journal bearings in locomotives among other things. Alloy 851 has the best casting characteristics of the three. Machinability is excellent for these alloys.

Alloy	Approximate Solidification Range F
208	1160-970
242	1175-990
295	1190-970
319	1220-960
332	1080-970
336	1050-1000
356	1135-1035
360	1105-1035
383	1080-960
390	1200-945
443	1170-1065
535	1165-1020
771	1185-1120
850	1200-435

The compositions of aluminum casting alloys are found at the end of the book

ALUMINUM WROUGHT ALLOYS:

I often mix wrought scrap with pistons or Briggs and Stratton parts to make a good general casting alloy. I may or may not add copper. The composition and application of a few wrought alloys is included below.

Wrought Alloy Numbering System

Aluminum 99.0% or greater	1xxx
Copper	2xxx
Manganese	3xxx
Silicon	4xxx
Magnesium	5xxx
Magnesium and Silicon	6xxx
Zinc	7xxx

Alloy	Common Applications
1xxx	Chemical resistant and electrical parts
2xxx	Aircraft and truck parts.
3xxx	Cooking utensils, tanks, mobile home parts
4xxx	Forged engine pistons
5xxx	Auto-body sheet
6xxx	Welded structures, architectural parts
7xxx	Airframe and highly stressed parts

Alloy	Al	Si	Cu	Mn	Mg	Cr	Ni	Zn
2014	93.5	0.8	4.4	0.8	0.5			
2024	93.5		4.4	0.6	1.5			
3003	98.6		0.12	1.2				
3004	97.8			1.2	1			
3105	99			0.55	0.5			
4032	85	12.2	0.9		1		0.9	
5182	95.2			0.35	4.5			
6061	97.9	0.6	0.28		1	0.2		
6063	98.9	0.4			0.7			
7050	89		2.3		2.3			7.6

MELT REACTIONS:

The strength of castings, especially aluminum alloys are limited by two types of defects: folded oxide films within the casting and stresses left in the casting from being quenched in water after heat treatment. We are concerned with casting, therefore oxide films will be discussed.

Liquid metal is a highly reactive material. It reacts with the gasses around it, the slag and other material it comes in contact with. Water, when in contact with molten metal, breaks down into hydrogen and oxygen that either dissolve into the metal or form oxides of the metal or one of the alloying components of the metal. Metals dissolve increasing amounts of gas with higher temperatures. As the metal cools, the gas comes out of solution and forms bubbles or pinholes in a casting. Hydrogen porosity is a common gas related casting defect.

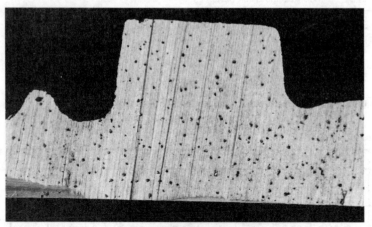

Hydrogen Porosity

This metal shows gas porosity from pouring into a damp ladle. It also had many entrained oxides from pouring into the ladle from sufficient height to cause turbulent flow.   1.4x

Many metals form oxide films on their surface as they come in contact with the oxygen in the air. A solid film quickly grows on the surface of aluminum. When the film remains on the surface, it is not harmful but provides a cover over the metal to prevent further oxidation. When pouring aluminum, it oxidizes so quickly that it forms an oxide tube around the stream. The oxide tube often breaks apart and goes down the sprue into the casting. Another tube is immediately formed and broken. If a pouring basin is properly used, the oxides may separate out and not go into the casting.

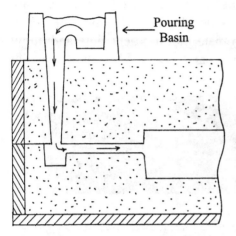

Pouring Basin

When pouring from a furnace into a ladle, the falling oxide tubes fold like an accordion on the surface of the metal in the ladle. If the falling metal stream is high enough, air is taken in along with the stream, rising as bubbles and leaving a trail of oxides in the metal. Each of these oxides submerged in a casting is a small crack.

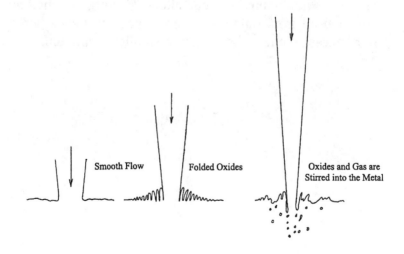

Smooth Flow   Folded Oxides   Oxides and Gas are Stirred into the Metal

33

The strength of a casting is limited if it is full of these small cracks. Therefore, for high strength or pressure-tight castings, the idea is to limit the turbulent flow that breaks apart the oxide film and stirs it into the metal. Anything that causes the molten metal stream to break up, jet or draw in air is to be avoided. One of the main functions of the gating system is to eliminate the turbulent flow and allow the oxides and slag to separate out of the molten metal flow.

Large Folded Oxides in an Aluminum Casting. 3x

Turbulent pouring is illustrated by pouring water into a pot from increasing heights. As the height increases, the sloshing and splashing become greater and air is taken under the surface of the water forming bubbles.

Iron dissolves in aluminum and is also a source of brittleness in aluminum silicon alloys. Coating iron melting pots and all iron tools that come in contact with a boron nitride wash reduces iron pickup by molten aluminum.

MELTING PRACTICE
MELTING ALUMINUM IN THE FOUNDRY:

When melting aluminum, the best results are obtained when the melting is rapid. Maintaining a molten heel of aluminum in the crucible speeds the melting process. If multiple charges are to be heated, it is good practice to leave 1/3 of the crucible filled with molten aluminum to reduce the thermal shock to the crucible and speed melting of the new charge. All tools, fluxes, pots, or anything that comes in contact with the molten metal must be preheated to minimize the water present on their surfaces, or in the flux. Keep the temperature of the aluminum as low as possible, only as high as needed to properly pour. Do not hold the metal at high temperatures for long periods of time. Melt quickly and pour.

Problems encountered when melting:

The two main problems encountered when melting aluminum are **dross formation** and **gas porosity**. Both the metal charge and the furnace should be clean when melting. The crucible and tools should be cleaned of caked on dross and metal by scraping after each heat. If this material is not removed, it could slough off into later melts and become mixed with the aluminum. Charge materials should be clean and free of oxides, water or oils. Oxides absorb moisture that can cause porous castings. The best way to eliminate moisture and oils is to preheat the charge materials.

Dross Formation:

Metallic aluminum is normally covered with a thin film of oxide because the metal readily oxidizes in air at room temperature. This oxide film forms a protective barrier against further oxidation. When melting aluminum, the exposed surface of the molten metal will oxidize to form dross. This dross may float on top, sink to the bottom, or it may become mixed in the melt. If the dross is not "wetted" by the aluminum it will float. However if it is wetted, it may become mixed into the melt or sink to the bottom. The specific gravity (weight relative to an equal volume of water) of some of the materials found in dross are listed below.

Specific Gravity of the Components of Dross	
Mg	1.74
Si	2.40
SiO_2	2.40
$Al_2O_3\ 3H_2O$	2.42
Al	2.70
MgO	3.65
Al_2O_3	3.99
CuO	6.40

Other factors that increase the amount of dross include the use of magnesium containing alloys in the charge material, using fine, thin or corroded scrap and high gas temperature at the melt surface. Rapid melting reduces the amount of dross formation by limiting the time the melt is exposed to the atmosphere. Melting losses are also higher with "soaking" or holding the metal at high temperature for an extended time and when fuels are burnt with a large excess of air. An unbroken film of oxide provides protection to the melt preventing further oxidation and absorption of hydrogen. However, if the film is broken, oxidation starts again at the break.

Gas Porosity:

Generally, liquids dissolve less gas at higher temperatures, however metals dissolve more gas at higher temperatures. Hydrogen easily dissolves into molten aluminum above the melting temperature. At 1220° F there is a large increase in the amount gas that can be dissolved in the melt and as the temperature increases, so does the ability

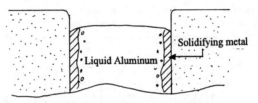

Hydrogen gas is ejected from the solidifying metal.

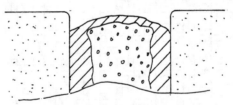

The top freezes over trapping the gas.

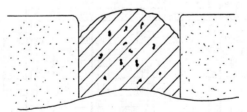

Gas pressure forces the top up forming a cauliflower head.

Gas in a riser

of the aluminum to dissolve more gas. As the casting cools, the ability of the aluminum to hold the gas in solution decreases and the gas forms bubbles. These appear as pinholes in finished castings. Gassy metal will often form a "cauliflower" head at the top of the risers.

Hydrogen in the aluminum melt comes primarily from water vapor. The gassing of aluminum due to water vapor is caused by the reaction:

$$2Al + 3H_2O \rightarrow AlO_3 \text{ (dross)} + 6H \text{ dissolved in the aluminum}$$

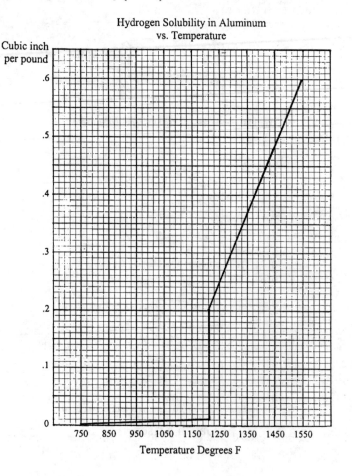

Hydrogen Solubility in Aluminum
vs. Temperature

Cubic inch per pound

Temperature Degrees F

The amount of water vapor required to affect the aluminum is so small that the amount of vapor in one cubic inch of air can ruin over 1 pound of metal.

Water vapor can come from the surface of the charge materials, oxides on the charge materials, damp flux, dirty tools and skimmers, or from the products of combustion. Hydrogen can also come from oily scrap. Fortunately, most of the water vapor and oils can be removed by preheating and keeping anything that comes in contact with the metal HOT. Hydrogen can also become trapped in the surface dross. If stirred, this can gas the melt.

Aluminum reduces many oxides. If any iron oxide (rust) is present, then the reaction below occurs.

$$Fe_2O_3 + 2Al \rightarrow Al_2O_3 + 2Fe$$

When in contact with siliceous materials, the following reaction takes place:

$$4Al + 3SiO_2 \rightarrow 3Si + 2Al_2O_3$$

The silicon goes into solution, and dross is formed.

Aluminum dissolves iron. Steel is more soluble than cast iron. However, chromium steels are the most resistant to liquid aluminum's corrosive action. The type of alloy melted also has an effect on molten aluminum's attack on steel with zinc containing alloys being the most corrosive. Washes are used over steel tools and crucibles to prevent pick up of iron. There are several commercial washes available, however many shop-home-brews containing mica, talc or limestone and sodium silicate are also used. One such wash consists of seven pounds of powdered limestone ($CaCO_3$) and 4oz sodium silicate mixed to a gallon of water. A warm ladle ($200°$ to $400°$ F) is painted so that the material sets quickly, and then it is dried at red heat. One ladle wash readily available to the small foundryman is Alum-a Kote sold by Mifco. Ladle wash coatings will have to be renewed as they crack when the ladle cools.

Clean foundry practice prevents most of the gassing problems, however there are degassing methods. Aluminum may be degassed by bubbling dry nitrogen, argon or a mixture of nitrogen and Freon, or nitrogen and chlorine or nitrogen and fluorine. Chlorine works well because it lessons the surface tension of the metal and allows the oxides to rise to the top. The best degassers make very fine bubbles that are well dispersed in the melt. These smaller bubbles do not cause much turbulence at the surface of the metal, so are less likely to stir contaminants back into the melt. This is accomplished by using a lance with fine holes or pores. Porous tips are common on lances. Chlorine degassing tablets are used by forcing them to the bottom of the ladle with a rod and cup. Some home foundry-men use a half teaspoon of powered "pool shock" wrapped up in a section of aluminum foil to degas their melt (keep it dry). I have not tried pool chlorine and can not comment on its effectiveness. The EPA has reduced the use of straight chlorine.

Commercial solid fluxes usually contain 3 ingredients: sodium chloride (table salt), potassium chloride, and small amounts (approximately 10%) of fluorides. They may also contain an oxidizing compound that generates heat to improve the flux wettability. Chloride salts melt and float on the top of the molten metal surface forming a barrier that prevents oxidation. The fluorides reduce the adhesion between the aluminum oxides and the liquid metal allowing the oxides to separate out. Cover fluxes have very small amounts of fluorides. Cleaning fluxes have larger amounts of fluorides and an oxidizing agent. Drossing fluxes are the most reactive fluxes because they have still larger quantities of fluorides and oxidizing compounds. Treated dross should lose its shiny metallic look and turn dark and powdery.

Often, unless properly dried and carefully used, fluxes often create more problems than they cure for the home foundryman. Some powered fluxes are added to the bottom of the crucible at the beginning of the melt. This allows them time to dry out before the aluminum melts. Degassing should not be used to remedy poor melting practice. Problems are minimal with "clean melting practice."

The pouring temperature of aluminum is approximately 1380° F. Pouring temperature is dictated by the fluidity of the particular alloy and the type of casting. Long thin castings require higher temperatures than thick chunky castings. Many castings work well between 1325° F and 1350° F. Some castings pour as low as 1250° to 1275° F.

Left-Testing the Temperature of the Molten Aluminum Using a Digital Thermometer

On the shop made thermometer used on the left, the digital thermometer is made by "Fluke" and the thermocouple tip is sold by "Mifco." It is good to 2400° F however, I have never had it above 1900 F.

III. COPPER ALLOYS, BRASS and BRONZE

Because copper may be alloyed with many elements, there are many copper alloys. Some alloys are specific casting alloys and others are primarily wrought alloys. Because some may choose to melt scrap, both wrought and casting alloys are briefly discussed.

High conductivity copper castings are used for both electrical parts and for thermal conductivity applications. Heat exchanger parts and water-cooled furnace tuyeres (air inlets) may be made of conductivity copper. Electrical and thermal conductivity are closely related. Because small additions of alloying elements greatly reduce the conductivity of copper, their amounts are closely controlled. Phosphorus drastically affects conductivity, therefore deoxidation with phosphorus is kept to a minimum in conductivity copper castings. See the graph on the next page for conductivity vs. impurities for copper alloys.

Brasses and bronzes have a wide range of properties and are popular casting alloys. They usually melt at temperatures below 1900° F, are corrosion resistant and are formulated in a variety of colors. Manganese bronze has high as-cast strength and is used for propellers. Aluminum bronzes have high strength and some aluminum bronzes are heat treatable for very high strengths.

Brass is an alloy of copper and zinc. It is usually made by first melting copper and adding zinc. Zinc boils at 1665°F. When added to molten copper, much of the zinc "flares" or boils off before the temperature can be lowered. As a result, it is difficult to match colors between different heats. Brass can be remelted with relatively small zinc losses. The color of copper-zinc alloys is approximately as follows:

98% copper - copper color

90% copper, 10% zinc – antique gold or dark bronze yellow

85 –80% copper, 15 to 20% zinc – red brass or copper color

70% copper, 30% zinc – bright yellow, yellow brass

60% copper – 40% zinc- light yellow, whiter yellow.

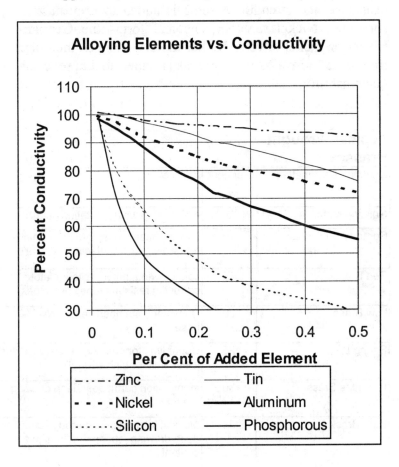

Depending on the zinc content, brasses are divided into three groups: the low brasses contain less than 20% zinc, the high brasses contain 20 to 39% zinc, and Muntz metal with 40 to 42% zinc. Muntz metal is also called "alpha-beta" brass. Increasing the zinc content over 32.5% produces a harder brass but reduces the ductility. The maximum zinc content is about 36% for the best combination of strength and ductility.

Other Additions to Brass: Tin, nickel, aluminum, silicon, manganese, iron and arsenic may be added in small quantities. Tin is added to brass to improve corrosion resistance and strength. Arsenic is added to prevent zinc corrosion. Nickel replaces zinc and forms the German-silvers or nickel silvers. Nickel rapidly changes the color from pale bronze to white. Lead is added to improve the machinability.

Typical Wrought Brasses

Common Designation	Per cent Zinc	Applications
Guilding Metal	5	emblems, jewelry, imitation gold
Commercial Bronze	10	jewelry, stampings, forgings, small hardware, screws, rivets
Red Brass	15	cold drawing, automobile radiators, corrosion resistant tubes and pipes
Low Brass	20	Drawn and stamped items, flexible hose
Spring Bass	25	Upper limit for corrosion resistance and lower limit for high strength
Cartridge Brass	30	ammunition cartridges, deep drawing
High Brass	34	Musical instruments, cartridge cases, deep drawing, cheap due to high zinc content
Muntz Metal	42	condenser tubes, architectural work, pipes

COPPER-BASED CASTING ALLOYS:

Not all copper-based alloys are well suited for the small foundry. Some are difficult to cast while others may generate more smoke than your neighbors are comfortable with. Although you might think yellow brass is a logical choice, it is difficult for a beginner to cast and zinc fumes may draw unwanted attention to your project.

The best copper based casting alloy for home casting is **silicon bronze** (95-1-4) . It may be melted in any type of furnace without loss or change in composition and produces no smoke or fumes. It may be repeatedly remelted. It is particularly well suited to cupola melting. It is easy to cast and takes intricate forms well. It forms little or no dross and does not require a flux. Silicon bronze castings are clean and free from sand. Castings are ductile, take a high polish, and may be patinaed. Silicon bronze is available in several colors and is less expensive than tin bronze. Castings are strong and may be welded.

Melt silicon bronze quickly under a slightly oxidizing atmosphere. Avoid holding the molten metal in the furnace and pour the molds as quickly as possible. The pouring temperature is between 1900° F and 2150 F, depending upon the size and thickness of the casting.

Use of a charcoal cover is not recommended because the alloy absorbs gas from the charcoal. Silicon bronze may be degassed with nitrogen or a flux. It usually does not need deoxidizing.

Silicon alloys must be separated from other copper-based alloys because silicon is extremely deleterious to other copper alloys. It combines with lead to form lead silicates causing a white scum and a wormy wrinkly surface.

Sand mixes for silicon bronze should be under 6% moisture to avoid dross formation and pinholes.

Gating and risering are very important. While silicon bronze does not shrink as much as manganese and aluminum bronze, it shrinks more than tin bronze.

Silicon bronze is sold under the trade names of Herculoy, copper 92%, zinc 4%, silicon 4% and Everdur, copper 95%, silicon 4% and manganese 1%. It comes in ½-inch cubes or 20-pound ingots.

Leaded Red Brass is typically copper 85%, tin 5%, lead 5%, and zinc 5%. Semi-red brass is copper 81%, tin 3%, lead 7% and zinc 9%. It is used in marine fittings, pumps, faucets, statues, and architectural castings. These alloys are easily melted in most furnaces. They are rapidly melted under an oxidizing atmosphere, removed from the furnace cooled in air, deoxidized and poured at the proper temperature. Red brass is deoxidized with 1 ½ oz 15% copper phosphorous shot per 100 lbs of melt. An excess of phosphorous makes the metal too fluid causing it to penetrate into the sand; making dirty castings. Zinc is added to replace what has burned out. General practice is 1 to 1 ½ pounds per 100 pounds of melt, however individual foundry experience is best used here.

Pouring temperature is 1900 to 2250° F. Clean charge material virtually eliminates the need for flux. Glass and borax may be used as a flux. As with all copper alloys, gating is important.

Yellow brass is copper 67%, tin 1%, lead 3%, and zinc 29%. It may also contain copper 63%, tin 1%, lead 1% and zinc 35%. It is found in ornamental castings and inexpensive fittings.

Yellow brass is much more difficult to cast than red brass however, it is melted much like red brass. There are higher melt losses, especially with thin scrap. Thick sections may be melted to form a molten heel in the crucible and the thinner sections are pushed under it to reduce the losses. Due to the excessive white zinc fumes, yellow brass is unsuitable for cupola melting. Red and yellow brasses are compatible and may be mixed.

Flux is generally not required because the zinc fumes prevent hydrogen pickup. Yellow brass does not require deoxidizing. Pouring temperature is 1900 to 2050° F with thicker sections and higher zinc content being towards the lower end. Higher temperature causes dirty castings due to zinc flaring inside the mold. The molds must be poured very quickly to prevent zinc oxide from forming in the mold. Slow pouring results in a wormy appearance. Molds may be tipped up with the gate at the lower part of the mold to prevent aspiration of air in the sprue. Gas also has a better chance to escape from an inclined mold with a vent at the top. Metal spreads over an inclined surface more gently than a flat surface reducing turbulence and minimizing oxide formations.

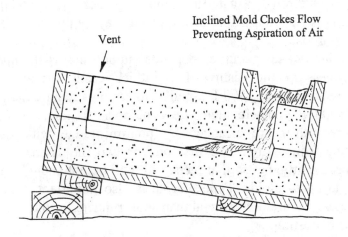

Inclined Mold Chokes Flow
Preventing Aspiration of Air

Vent

Runners for yellow brass are proportioned with 1/3rd in the drag and 2/3rds in the cope. This allows the sprue to choke quickly.

Manganese Bronze or high strength yellow brass typically contains 64% copper, 26% zinc, 4% manganese,

3%iron and 3% aluminum. It is very strong and has good corrosion resistance. Manganese bronze is used for propellers and non-sparking tools. They cast fairly well and have a good finish even in coarse sand due to the protective aluminum oxide film carried on its surface. Manganese bronze has high shrinkage therefore gating and risering are very important. Unusually large risers may be required. Chills may be required for thick sections to set up directional solidification. Manganese bronze is usually melted in a crucible furnace.

Pouring at a uniform speed while keeping the sprue filled is important. Oxides will float in the pouring basin or on top of the sprue, this will keep any oxide that have pastsed the skimmer from going down into the casting. Hot metal must be fed into the risers to take care of the shrinkage. If it is not possible to introduce hot metal into the riser through a gate, they must be back fed with hot metal. An extra crucible of hot metal may be provided for this purpose.

Pouring temperature depends upon the size and thickness of the casting. Large thick castings may be poured as low as 1800 F. Small thin castings require metal at 1900 F. Where there are a number of molds to be poured, they are arranged by pouring range with the thinnest ones first. The first one or two molds may be poured a little hot so that metal will still be hot enough for the last few molds. Type of gate is also important. Long runners require hotter metal than short runners for the same type of casting.

When gating, the metal must enter the mold without splashing, squirting or turbulence of any kind. Any turbulence in the sprue, runners or mold will cause dross and a rejected casting. Strainer cores break up the flow and create dross in manganese bronze therefore are not used. Runners are often lined with nails to catch any dross before it enters the mold cavity. Gates should have dead ends to

48

catch the first metal in the mold. The proper size of gates and risers is only gained from experience. Sometimes the mold is rotated from horizontal to vertical to bring the riser directly over the section to be fed.

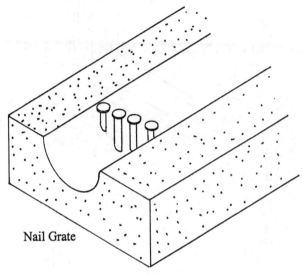

Nail Grate

The molds should be filled by a simple displacement from the bottom. A reverse horn gate may be used. A horn gate may be made as a dry sand core that is rammed up and left in place.

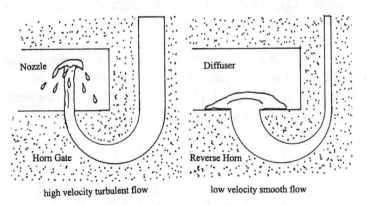

Nozzle   Diffuser

Horn Gate   Reverse Horn

high velocity turbulent flow   low velocity smooth flow

High velocity turbulent flow is undesirable in dross forming alloys.

49

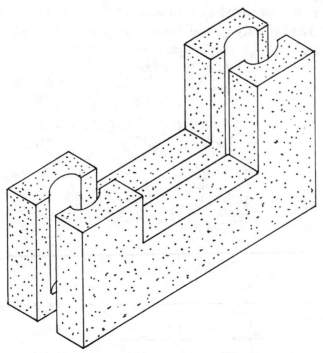

Bottom gate made from core sand and glued together.

Removal of gates and risers from manganese bronze castings is difficult. Various neck down cores have been used on risers to help alleviate the problem.

Tin Bronze is often used for pressure castings such as valves and pumps. It is also used for marine hardware and steam fittings. The most common type is gunmetal, copper 88%, tin 10% and zinc 2%; however there is also leaded tin bronze. **Bell bronze** contains about 20% tin with varying amounts of zinc and lead. Because of the high tin content, bell bronze is brittle and cracks easily. Extra care must be taken in casting and annealing.

Tin bronze is easy to cast and free from the drossing problems of manganese bronze. It exhibits less shrinkage than manganese bronze but still requires large risers. Melt with an oxidizing atmosphere and deoxidize the 2 ounces of 15% phosphorous shot per 100 pounds of melt. Pouring

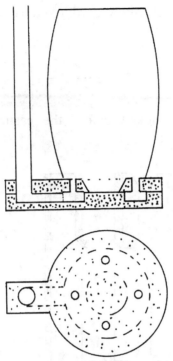

By using two small dry sand cores with multiple gates, you may gain
advantage of bottom gating and avoid the jet from a single horn gate.

temperature is 1950° to 2250° F. Molds are occasionally
dusted with wheat flour and graphite.

High Lead Tin Bronze Bushing Alloys. SAE 660
copper 83%, tin 7%, lead 7%, zinc 3%, and SAE 64, copper
80%, tin %10, and lead 10% are common alloys. SAE 64 is
probably the most popular bearing bronze. The high lead
content of SAE 64 produces a characteristic gray-black
patina making it a good **statuary bronze**.

High lead alloys easily penetrate all but the finest sand.
Fine sand facings, washes or additions of sea coal are used
to minimize the metal penetration.

Bearing alloys, because of their fluidity and weight,
have a tendency to wash sand into the mold. Inverted horn
gates, baked sand sprues, runners and ingates reduces this

washing tendency. Rough or eroded sprues and runners indicate sand has probably washed into the casting.

Gates should be choked before the casting to reduce nozzling or spraying.

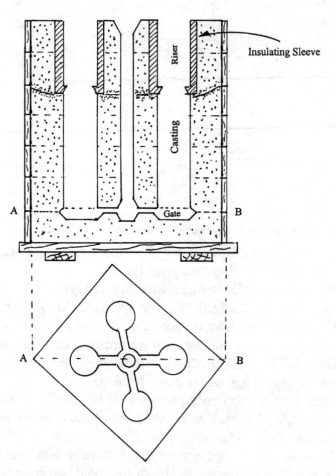

Using stacked molds to cast bronze round stock.

52

Lead Sweat is the separation of liquid lead from a solidifying casting. This lead may appear on the surface of the casting as drops or it may sink to the bottom of the casting. In most copper-lead alloys, because lead has a solidification temperature of 621° F, it remains as droplets after the rest of the alloy has solidified. Shaking out a high lead casting before the lead has solidified, may result in excessive lead sweat and porous castings. Because lead precipitates late in the freezing, it may fill areas that otherwise might have become porosity. Lead in copper alloys often makes it easier to produce pressure-tight or leakproof castings.

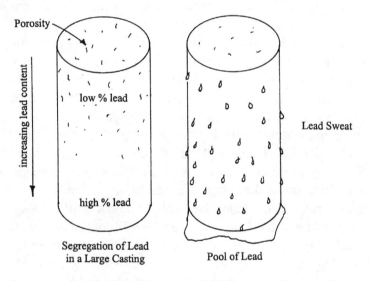

Segregation of Lead in a Large Casting

Pool of Lead

During solidification, copper crystals separate from the liquid alloy until the lead concentration in the remaining liquid reaches 41%. At this point, a heavy new liquid is formed that is 92.6% lead and sinks to the bottom. In smaller castings where the cooling is fairly rapid, dendrites grow quickly trapping the lead droplets as a finely divided phase. The longer cooling times of larger castings allow more lead to segregate to the bottom of a casting. The top of the casting may have a much lower lead concentration than the bottom resulting in a spongy top with large lead blemishes at the bottom. In such cases, the sprue is frozen off with a chill and the mold is turned over as it cools allowing lead to flow between both ends of the casting.

Aluminum Bronze is difficult to cast because of its high tendency for foaming and drossing. It has a short solidification range and therefore has extreme piping. Aluminum bronze requires the use of large risers or chills and insulating sleeves. Due to the thin film of aluminum oxide formed during pouring, better surface finish will be found on aluminum bronze castings than tin bronze castings in the same sand. Aluminum bronze is used for bearings, ball-bearing races, non-sparking tools, pump bodies, heavy duty gears, and wire drawing dies.

Aluminum bronzes are defined as copper alloys containing 5 to 15% aluminum, up to 10% iron with or with out manganese or nickel. Alloys of 9 to 11.5% aluminum are heat treatable for an increase in hardness with a corresponding loss in ductility. Castings of more than 9% aluminum become brittle when slow cooled in sand molds and must be shaken out before the temperature reaches 1050 F; preferably as soon as they are cool enough to handle without distortion. The addition of more than 2% nickel reduces the tendency to precipitate a brittle phase.

Melting should be quick, as for all copper alloys. The melt may be gassed from reducing furnace atmospheres. Stirring or agitation of the melt in an oxidizing atmosphere results in heavy drossing of the aluminum and high losses.

Pouring temperature is between 1950° and 2250° F. The ladle should be well skimmed and poured as close to the mold as possible to reduce drossing.

Moisture should be low in green sand molds to prevent pinholes. Moisture also contributes to the formation of surface dross. Facing sands are generally not used and a more open or permeable sand is preferred, especially for large flat castings. Gating must introduce the metal quietly at the bottom of the mold. Squirting of the metal causes drossing and must be avoided.

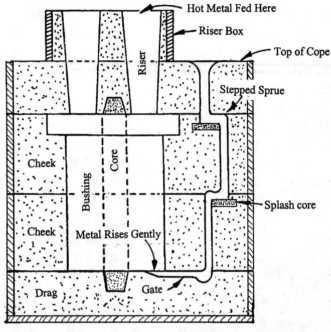

Stepped Sprue and Splash Cores Slow Metal for Gentle Flow

Suggested gating for an aluminum bronze bushing.
Note: Chill the sprue and fill risers. See page 113

Casting Aluminum Bronze Spheres:

Solid aluminum bronze spheres, used in ball valves, are produced in green sand molds by careful foundry practice.

The spheres, machined to 3.5 and 4.5 inches in diameter, are cast from ingots with an analysis of 87% copper, 10.5% aluminum and 2.5% iron.

The pouring temperature is kept as low as possible at which complete castings can be regularly produced. In consideration of the heavy cast iron chills used in the molds, the ball valves are poured at 2150°F. Other bronze in the shop is poured between 1950° and 2050°F.

In order to reduce folded seams of oxides in the castings, the molds are gassed using carbon dioxide. Because it is heavier than air, carbon dioxide displaces the air. A lit match is held over the riser openings as gas flows into the mold through a rubber hose. When the mold is full the flame is extinguished. The casting is immediately poured. Because there is no free oxygen in the mold, no oxide is formed.

Although the ball castings look simple, much time and study were required to develop a process that would assure good castings. The two main imperfections were shrinkage and dross. To overcome the shrinkage, cast iron chills approximately 1-inch thick are rammed up in the drag half of the mold.

Because the chills must be free of any scale, rust or moisture, they are sandblasted before each use. The chills are preheated to prevent cold shunts in the castings. A rough but sufficient test of the temperature is made by hand. If one can barely touch them, they are at the proper temperature. If they are too hot, they dry the sand at the bottom of the mold causing washes. They also do not function well as a chill being too hot. The chills are preheated by placing hot chunky pieces of steel in them.

The mold is closed and quickly poured to prevent any moisture from forming in the mold.

A sprue leads to a runner that is split between the cope and the drag. This runner design catches dross in the cope. The horn gate is attached to the runner by a short gate that is cut in the drag only. An 8-inch cope is used and a 2-inch diameter riser is placed directly over the casting. Two balls, with a combined machined weight of 9.5 pounds, requires 32 pounds of metal to be poured.

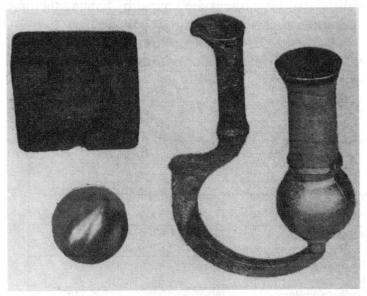

The larger spheres are poured as single castings. Each ball weighs 11 pounds and requires 33 pounds of metal. The 4.5-inch ball requires 2 chills, one on each side. Originally, the chills extended 1 inch up the riser neck. However, this caused the riser to solidify before the liquid center of the ball, causing a shrink on the top of the casting. The problem was corrected by reducing the chilled section to about .5-inch.

The ingate through the bottom of the casting also gave trouble. If it was too long, it froze before the mold filled. If it was too narrow, it caused the metal to spout and if too large, it caused a shrink in the finished casting. The final horn gate, of square cross section, is seen in photo 2.

It is most critical to maintain a steady stream of metal when pouring. The flow is controlled by looking down into the riser. If the metal enters with a fountain like stream, the casting is defective. One single splutter is enough to cause sufficient foaming to ruin the casting. When looking down into the riser, the metal must rise with a perfectly smooth surface without a single ripple on it. To limit the oxidation of the metal, the castings are poured by holding the lip of the crucible as close to the sprue as possible.

Solidification Range of Copper Based Alloys		
Silicon Bronze	95-1-4	1680-1510
Leaded Red Brass	85-5-5-5	1850-1570
Semi Red Brass	81-3-7-9	1940-1540
Yellow Brass	67-1-3-29	1725-1700
Yellow Brass	63-1-1-35	1688-1657
Manganese Bronze	64Cu-26Zn-3Fe-3Al-4Mn	1693-1625
Tin Bronze	88-10-2	1830-1570
Tin-Bushing Bronze	80-10Sn-10Pb	1705-1403
Tin-Bushing Bronze	83-7Sn-7Pb-3Zn	1790-1570
Aluminum Bronze	88-3Fe-9Al	1915-1905

Temperature Difference, Beginning to End of Solidification.

Short Freezing Range Alloys	90°F or less
Intermediate Freezing Range Alloys	90 to 200°F
Long Freezing Range	Over 200°F

BRASS & BRONZE FOUNDRY PRACTICE:

Copper-based alloys are cast in green sand, baked sand, cement bonded molds and plaster molds. Large castings are made in cement bonded molds. Plaster molds are used for small intricate castings, hardware and ornamental castings. Leaded alloys do not cast as well in plaster molds because lead reacts with calcium sulfate and discolors the finish. Lead should be minimal in the alloys cast in plaster molds. The majority of brass and bronze castings are made in green sand molds.

MOLDING SANDS:

Because many copper alloy castings require a good "as cast" surface finish, fine molding sands from AFS 120 to 270 with good flowability are required. Fine, lower permeability sands give a better finish. Fine synthetic sand may be bonded with clay, however if it does not have good flowability, it will not pack well during molding and produce a poor surface finish. Naturally bonded sands with higher percentages of fines produce good results. French sand with a grain fineness of 176 was considered the best natural sand for statuary work and fine art. Albany sand is commonly used. Yellow Reba produced in Surrey, British Columbia also produces fine work*.

As the casting size increases, more gas is produced in the mold and the permeability of the sand must be higher. Casting sands are specified by casting weight. Although green sand is used for castings over 1000 pounds, dry or baked sand molds give consistently better results for castings over 50 pounds. Green sand molds use a maximum of 6% water in order to prevent excessive drossing and gas cavities. Water content may be as low as 3%. Copper-based alloys are easily cast in Petrobond sand because of the fine grain used and the absence of water.

*Stewart Marshall's Bronze Formula

CORE SAND:

In order to prevent hot tears, collapsibility is important in core sands used for copper-based alloys. The core sands must be fairly low in hot strength so that they collapse quickly as the casting shrinks. Addition of wood flour or fine sawdust, additions of vents and lower amounts of binder all increase the collapsibility of cores. Hollow cores collapse readily in copper alloy molds. Depending upon the alloy being cast, core washes may or may not be used. Aluminum bronze is less susceptible to hot tearing so collapsibility is less important for this alloy.

When casting thick internal sections that can not be easily fed, cores may be made of a chilling sand such as zircon or chromite.

GATING:

Copper-based alloys form dross because they contain elements such as zinc, aluminum and tin that form oxides. Because they are lighter than the molten alloy, these oxides tend to float out rather than become entrained as seen in aluminum alloys. Dross is most likely to become entrained in high zinc brasses and aluminum bronze. Gating arrangements that catch dross are used for copper alloys. In aluminum practice, metal flow is controlled at the base of the sprue; however gating of copper-based alloys is similar to that of cast iron. Metal flow in cast iron practice is controlled at the ingates or at the runners just ahead of the risers. Sprue to runner to gate areas of 2:8:1 or 3:9:1 allows the dross to float out in the runner. Because dross will float out in a riser, the runner may be made much smaller and a ratio of 1:2 may be used, when gating through a riser. Reverse gating, bottom gating, skim cores and traps or slag pockets are used to prevent dross from entering the casting. Any turbulence, squirting or agitation of the molten metal increases dross formation. Because

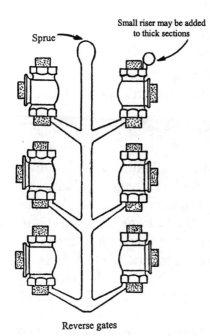

Sprue

Small riser may be added to thick sections

Reverse gates

strainer cores break up the flow into many fine streams, exposing the metal to increased oxidation, they are avoided.

Copper-based alloys are similar in density to iron for which formulas for gating and pouring exist. They may be used to estimate flow rates however *practical experience* is the only way to determine proper gating and produce good castings.

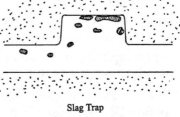

Slag Trap

A small shrinkage at the point where the gate joins the casting is a common defect on chunky castings. Raising the pouring head a few inches or changing the runner to the cope may solve this problem. On other castings a small riser on top of the gate next to the casting prevents this problem.

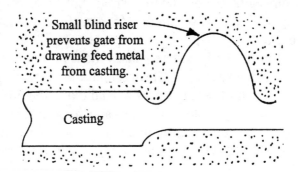

Small blind riser prevents gate from drawing feed metal from casting.

Casting

MELTING AND POURING:

Copper melts at 1982° F and aluminum at 1215° F. Although the temperatures used in copper-based alloys are higher than aluminum, aluminum requires about 60% more heat input or Btu per pound.

Melting and pouring operations greatly affect the quality of copper-based castings. Generally, copper alloys are best when melted in crucible furnaces. Although all alloys may be melted in a cupola, some alloys are particularly well suited to cupola melting, while others experience high zinc losses (yellow brass, manganese bronze). Because of the shallow bath (depth of metal) required and thickness of the slag cover, reverberatory furnaces are poor melters of copper alloys. Rotary furnaces, because the refractory underneath the melt is constantly being reheated, work well for copper-based alloys.

In order to minimize gas absorption, copper based alloys should be melted quickly and held at pouring temperature as short a time as possible. The metal should only be heated to pouring temperature plus 50 to100 degrees. Temperature should be measured with a thermocouple.

Because of the zinc, aluminum and tin contained in most copper alloys, drossing is a problem. Depending on the alloy, a flux of charcoal or a mixture of crushed bottle glass and borax may be used to cover the melt and prevent oxidation. The addition of borax makes a fluid glass flux. Minimum agitation or stirring while molten also reduces drossing and gas absorption. Manganese and aluminum bronze are particularly sensitive to drossing. The slightest agitation in the molten state causes them to foam with the foam solidifying as dross. Clean foundry practice as used in aluminum melting reduces gassing and oxidation problems.

The solubility of both oxygen and hydrogen increases rapidly with increasing temperature above the melting point. Since both hydrogen and oxygen may be present in copper alloys, water vapor may be formed therefore the gasses are somewhat self-regulating. Melting under oxidizing conditions reduces the hydrogen content of the melt. Hydrogen may also be removed by flushing with nitrogen as seen in aluminum practice.

Zinc flaring will reduce the hydrogen content of any high zinc copper alloy such as yellow brass or manganese bronze. Zinc flaring occurs when the zinc boils off and the vapor reacts with the oxygen in the air. Flaring is often used as a gage of proper pouring temperature. Because of the flaring losses, zinc is usually replaced at the rate of 1 to 1.5 pound per 100 pounds of melt. The actual amount is determined by practical experience.

Carbon and sulfur may also be present in the melt and form CO and SO_2 causing gas hole defects. Deoxidation minimizes porosity due to CO and SO_2. Phosphorus is added as copper-phosphorous shot to deoxidize the melt.

Copper-based alloys are usually melted in an oxidizing atmosphere and deoxidized with phosphorous before pouring. Burning fuel with excess air produces an oxidizing atmosphere. Burning with deficient air produces a reducing atmosphere.

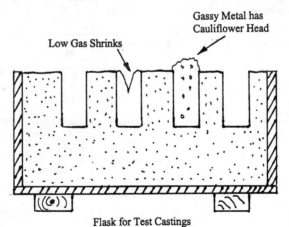

Flask for Test Castings

Test castings 1 to 2-inches in diameter are often made to check the quality of the metal. If the casting develops a puffy head (cauliflower head as seen in aluminum), then the metal is gassy. If the metal develops a deep pipe, the metal has very low gas content. Make test castings by rammming up a drag and using a 1 to 2-inch diameter steel tubing as found at muffler repair shops, or used as steel fence post, cut several holes 2 to 3-inches deep as seen in the drawing. Baked sand molds may also be made ahead of time and kept on hand.

Defects found by too high a pouring temperature include gassy metal, excessive drossing, porosity due to increased shrinkage and excessive zinc flaring inside the mold. Defects from too low a pouring temperature also include misruns caused by the metal freezing before it completely fills the mold. Cold metal does not give a high head or risers time to function, contributing to shrinkage and porosity.

Summary of Melting Data for Copper-based Alloys

Type of Alloy	Temperature in Furnace	Pouring Temperature	Flux if needed	Deoxidizer
Leaded Red Brass	2050-2300	1950-2250	glass and borax	1 oz / 100lbs 15% P-Cu
Leaded Yellow Brass	2050-2100	2050 max	glass and borax	P as needed
Manganese Bronze	1850-1950 flaring temp	1900 max	glass and borax, charcoal	if used: 1 oz / 100lbs 15% P-Cu
Tin Bronze	2050-2300	1950-2250	glass and borax, lime and soda ash, charcoal	2oz / 100lbs P-Cu
Silicon Bronze	pouring temp +50-100F	1900-2150	glass and borax	not used
Aluminum Bronze	pouring temp +50-100F	1950-2250	not used	not used

IV. BASIC METALLURY OF IRON:

There are several types of iron, **white iron**, **gray iron**, **ductile iron**, and **malleable iron**, being the main categories. **Mottled iron** is white iron that has small sections of gray iron in the center of the casting.

The properties of iron are largely dependent upon the amount and type of carbon in the iron. White iron has the carbon in combined form "iron carbide". It is very hard, brittle and has a white fracture when broken. Gray iron has the carbon in flake graphite form. It is soft and easily machined. It has a gray fracture. Ductile iron has the graphite in spherical form. Ductile iron may be bent without breaking. It has a steely surface when fractured. Malleable iron is made by annealing white iron so that the graphite separates from solution and forms spheres. It may also be bent without breaking.

The properties of iron vary widely. Iron may be extremely hard, soft, ductile (able to be bent) or have no ductility at all. Again, all these properties are related to the type and amount of carbon in the iron. Generally, steel has a maximum of 1.5% carbon. As the carbon is increased above 1.5%, it approaches semi-steel and as the carbon approaches 2%, it is called iron. There is no clear line that separates steel from iron. Low carbon steel, such as 1018 contains 0.18% carbon and is easily bent; however, as the carbon content is increased, the steel becomes harder. Medium carbon has 0.35% carbon and high carbon steel has over 0.45%. Very high carbon steel has .75% to 1.5% carbon.

Up to a point, the more carbon that is dissolved or combined in the iron, the harder it will be. Iron carbide, which is very hard and brittle, has the all of the carbon in the combined form. Once the carbon content reaches about 2.2 to 2.5 percent, it may start to separate out of the iron as graphite. Many properties of the iron are related to the type and amount of graphite that precipitates from the iron. Flake graphite is found in gray iron and spherical graphite is found in ductile and malleable iron. There are five types of flake graphite; however, they all cause the iron to be brittle and fracture as opposed to being ductile.

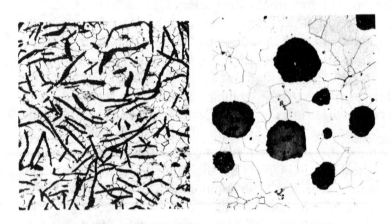

Flake Graphite and Spherical Graphite 250x
Courtesy International Nickel Co.

Carbon Content of Steel:

Carbon Class	Carbon range %	Typical uses
Low	.05 - .15	Chain, nails, pipe, screws, wire
Medium	.15 - .30	Structural steel, plate, bars
	.30 - .45	Connecting rods, axles, shafting
High	.45 - .60	Crankshafts, scraper blades
	.60 - .75	Automobile springs, band saw blades, anvils, drop hammer dies
Very High	.75 - .90	Chisels, punches
	.90 - 1.00	Knives, shear blades, springs
	1.00 - 1.10	Taps, dies, milling cutters
	1.10 - 1.20	Lathe tools, woodworking tools
	1.20 - 1.30	Files, reamers
	1.30 - 1.40	Wire drawing dies
	1.40 - 1.50	Metal cutting saws

Chemical Composition of Cast irons

Element	Gray iron %	White Iron-malleable iron %	High Strength gray iron %	Nodular Iron %
Carbon	2.5 - 4.0	1.8 - 3.6	2.8 - 3.3	3.0 - 4.0
Silicon	1.0 - 3.0	.5 -1.9	1.4 -2.0	1.8 - 2.8
Manganese	.4- 1.0	.25 -.80	.5 - .8	.15 - .90
Sulfur	.05 - .25	.06 - .20	.12 max	.03 max
Phosphorus	.05 - .1	.06 - .18	.15 max	.10 max

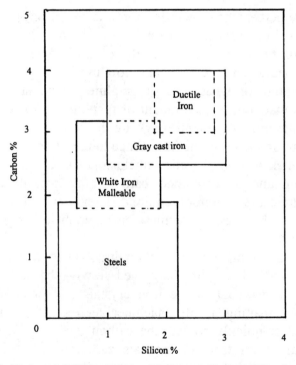

The Carbon and Silicon Relationship between Steels and Cast Irons

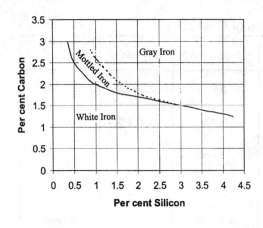

Malleable and ductile iron, both having spherical graphite, resemble steel in their ability to be bent. Hardness and hardenability are related to the amount of combined carbon Fe_3C.

CARBON:

The percentage of carbon in iron is called **total carbon.** The total carbon does not really describe the properties of the iron because it does not tell you how much carbon is combined as carbide Fe_3C and how much has precipitated as graphite or what type of graphite. There are several things that control the amount of graphite that precipitates from the iron. The main things are the cooling rate and the amount of silicon in the iron. Since thin sections cool faster than thick sections, "section thickness" has a lot to do with the amount of combined carbon and graphite. Thick machine parts, therefore require different chemistries to get the same hardness or softness and amount of graphite as thin parts.

At high temperatures, carbon dissolves in the iron up to about 4.5%. If the iron is cooled slowly, the carbon will work its way out of the iron crystals and precipitate as graphite making a soft, machineable iron. If the iron is chilled or quickly cooled, the carbon does not have time separate from the iron crystals and you get "white or chilled" iron that is extremely hard and brittle. Often, both properties are needed in a part, so a single surface may be chilled. Such is the case in train car wheels where the rim is chilled to produce a hard surface for wear resistance and the inner section is left soft for machining. The lobes of a camshaft are another example of a hard iron surface over soft iron. This heating and cooling effect is the basis for heat treatment of iron. Lathe beds cast in soft gray iron are flame hardened by heating the ways to dissolve the carbon back into the iron and then they are quickly cooled so that the surfaces are hard. Hard castings are softened by annealing or heating them up to the transformation temperature where the carbon is free to move, then slowly cooling them.

Because of the effect of thickness on cooling rate, a minimum section thickness is recommended to avoid chilled castings and misruns.

Suggested Minimum Wall Thickness for Iron Castings:*

ASTM Class of Iron	Minimum Wall Thickness, inches
20	.125
25	.25
30	.375
40	.5
50	.625
60	.75

*adapted from the Metals Handbook

Graphite has a low specific gravity of 2.25 so every 1% of graphite by weight represents 3.2% of the volume in the iron. Carbon will form 15 times its percent weight in iron carbide, Fe_3C. A white iron with 2.5% C will contain about 37.5% carbide (2.5% x 15 = 37.5%).

Silicon, sulfur, phosphorous, and manganese all affect the properties of iron. **Silicon** promotes graphitization (causes graphite to form) in cast iron. Silicon is used in gray iron from about 1% to 3.5% by weight.

Carbon equivalent describes the relationship of a particular iron to the eutectic point. The eutectic point is the lowest temperature a metal can be liquid. Changing the composition of the metal changes the freezing point of the metal, much like the addition of antifreeze or alcohol lowers the freezing point of water. If a metal has a eutectic composition, then it also has the lowest freezing temperature. Iron with 4.3% carbon has the lowest freezing temperature and is therefore the eutectic alloy. If the carbon content is higher than 4.3%, then carbon starts to separate out before the liquid iron freezes. If the carbon content is lower than 4.3%, then iron starts to solidify before the carbon. At exactly 4.3% carbon, both the iron and graphite solidify at the same time.

The carbon equivalent of an iron is represented by the equation:

CE = %Carbon in the iron + 1 / 3 %Silicon*

* sometimes the equation is modified by adding 1/3 phosphorous

CE = carbon equivalent

Calculate the carbon equivalent for an iron that has 3% carbon and 2.8% silicon.

$CE = 3 + (2.8 \div 3) = 3.93\%$

The carbon equivalent is 3.93%

If the carbon equivalent of an iron is greater than 4.3%, then graphite forms first during cooling. This graphite floats on the surface or pops into the air as sparkly flakes. This graphite is called kish.

Silicon is added to the melt as pig iron, or ferro silicon is added to the ladle. Silicon reduces the iron's ability to absorb carbon. In order to get high carbon iron, the charge material should be melted in the presence of carbon, then the silicon should be added to the ladle.

Phosphorous increases the fluidity of iron. It forms iron phosphide which is hard and brittle. The addition of phosphorus may lower the final temperature of solidification to about 1800 ° F. Phosphorous is usually limited to 0.3%. Small foundry operators may add the copper phosphorous shot used to deoxidize bronze, if they are looking for a fluid metal for thin castings. Piston rings usually have about .5% phosphorous for good fluidity.

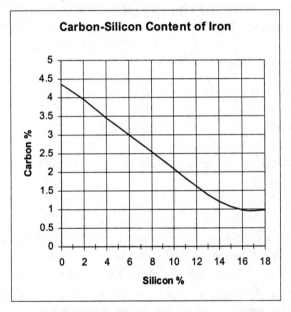

As the silicon content increases, the iron's ability to dissolve carbon decreases

Sulfur prevents graphitization or is carbide stabilizing. Low sulfur iron, 0.01% S will graphitize completely. Ductile and malleable iron require low levels of sulfur to properly form graphite spheroids. Sulfur is removed from the iron with alkali fluxes such as soda ash, lime, and calcium carbide.

71

Sulfur combines with iron to form iron sulfide. It separates out and goes to the grain boundaries as the iron cools. Because of its low melting point, iron sulfide remains liquid as the iron freezes. This liquid at the grain boundaries allows movement as the casting cools. This movement sets up large stresses in thin castings. Large thin gray iron castings may be susceptible to cracking, if the sulfur content is too high. Manganese is added to iron to reduce the sulfur content. This ties up the sulfur as manganese sulfide, preventing it from forming iron sulfide. To prevent stress cracking in thin gray iron castings, sulfur is kept below .12%.

ALLOYING ELEMENTS:

Copper is used up to 3% to add corrosion resistance and increase wear resistance in cylinder liners and brake drums.

Silicon: 13% to 18% silicon is used for acid resistant irons used in pipe fittings and pump housings.

Chromium is carbide stabilizing. It makes a hard and strong iron.

Nickel is slightly graphitizing and increases chemical resistance. Nickel irons are the toughest of gray irons in that they are able to absorb a lot of energy with out failure. They have a lower tensile strength between 20,000 and 40,000 psi, excellent machinability and good casting properties. Generally nickel additions of 3% are used to give a fine grain and the above mentioned properties. When added to ductile iron, the properties are considerably improved and the ductility is improved. Nickel 4.5% combined with chromium 1.5% makes a very hard, wear resistant iron known as "Ni-hard." Nickel may be added as pig in a cupola or as shot to the ladle. Nickel pigs are usually 3 to 3.5 pounds containing 96% nickel. Several other nickel pigs are available. Nickel recovery is almost 100% in cupola melting.

Right: These 1936 Harley Davidson Motorcycle Heads were produced by the Motor Castings Company of Milwaukee, Wisconsin. The iron contains 1% nickel and .3% chromium. The nickel improves the toughness and the thin fins are cast free from chilling or porosity.

Manganese: The main function of manganese is to neutralize the effects of sulfur. Sulfur forms a brittle iron sulfide. Usually manganese levels are about 5 times the level of the sulfur. Therefore, if the sulfur content is 0.1%, then the manganese should be about 0.5%. If the manganese reaches 1% then it has a carbide stabilizing effect. It may be added in a cupola charge as pig or silicomanganese briquettes. Manganese steel may also be added to the cupola charge. Silicomanganese or carbon ferro manganese may be added to the ladle. Gas porosity may be problematic with large ladle additions, especially with lower pouring temperatures.

DUCTILE IRON:

Ductile iron is made by adding magnesium to the ladle. This has the effect of deoxidizing the iron as well as reducing the sulfur. Magnesium also causes the graphite to form spheres rather than flakes. Because much magnesium is consumed by the sulfur, levels are reduced by using lime,

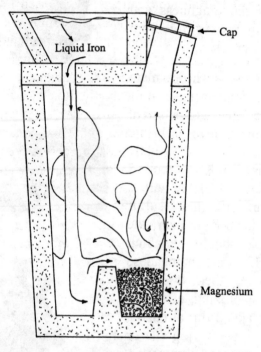

Treatment Ladle for making Ductile Iron

soda ash or sodium hydroxide as a fluxing agent before the addition of magnesium. Magnesium is usually in the form of magnesium ferro silicon. It may be added in a treatment ladle or placed in a refractory cup and plunged to the bottom of the ladle. The carbon content of ductile irons is between 3% and 4%. The silicon content is between 1.8% and 2.8%.

ALKALI FLUXES:

Sodium oxide is the active ingredient in desulfurizing fluxes such as soda ash (sodium carbonate Na_2CO_3) or sodium hydroxide (lye) NaOH. Soda ash contains 58% sodium oxide and sodium hydroxide contains 76%. Sodium hydroxide readily absorbs water from the air and must be stored in airtight containers and dried before use. Use of a wet flux may cause an explosion. Sodium hydroxide is highly corrosive and must be handled with gloves and goggles. It produces a large volume of fumes that burn the skin, throat and eyes. Adequate ventilation and protection are required for its use. It is used where very low sulfur levels are required. When in contact with molten iron, it breaks down into sodium oxide and steam. Sodium hydroxide is sold as "lye" at most grocery stores. Sodium carbonate (soda ash) is used to desulfurize iron and is used much more frequently than sodium hydroxide. It was sold as "Purite tablets" however now you may find it as a granular product from FMC corp. It is sold in 50 pound bags. Soda ash may also be found at pool supply stores, however it is usually in the powered form which is less desirable for cupola work. Soda ash is often used in water softeners and may also be purchased as "Arm & Hammer Washing Soda" at most grocery stores. Like sodium hydroxide, soda ash must be dry to prevent ladle explosions. It may be dried in an oven for a few hours before use.

When in contact with molten iron, soda ash breaks down into sodium oxide and carbon dioxide. The gasses generated by both alkali reactions stir the metal and increase their desulfurizing action.

The reactions between the flux and the sulfur are reversible, meaning that the flux will give sulfur back to the iron in certain conditions. The reverse reaction occurs when the temperature gets lower and there is little free flux left in the bath.

In use, the flux should be placed in a preheated ladle before the iron is tapped in. The reaction produces much gas and a layer of slag over the metal. Teapot type (bottom pouring) ladles are best for desulfurizing. Soda ash may also be added with the limestone in the cupola charge. It is a strong basic flux and slag liquefier. If you are melting in a crucible furnace, the iron charge should be layered with charcoal and flux. Generally, soda ash is added at the rate of .6 to 1% of the weight of the metal to get about 50% reduction in sulfur.

Powdered soda ash can not be added to the cupola charge when melting because it is too fine and blows out of the stack when the blast is put on. It is best to add it to a heated ladle. The soda ash makes a very watery slag that may be removed by adding "slag off" before pouring. Slag off causes the slag to stick in large clumps that may be easily removed.

Calcium Carbide (CaC_2) is also used to reduce the sulfur in cast iron. It is loaded into a graphite lance and injected into the melt with nitrogen.

Because ladle desulferization takes time, it is more difficult in a small foundry because the smaller ladles cool quickly. Larger ladles are usually heated.

V. GATING SYSTEMS

PARTS OF A GATING SYSTEM:

A gating system channels the molten metal from the top of the mold to the mold cavity or the casting. As the metal enters from the top, it goes into a **pouring cup** or **pouring basin**. Cups are molded or cut into the top of the sprue.

Molding a Pouring Cup. Notice the Steep Angle.

Smaller or less critical castings may use a cup but to minimize splashing and turbulence, a pouring basin is used. From the basin or cup, the metal enters the **sprue**. At the base of the sprue there is a well or enlarged area. At the base of the sprue there may be a **choke** that restricts and regulates the flow of metal into the **runner**. The choke may also be located at the entrance to the runner. The runner carries the metal to the **ingates** that open into the mold cavity. Sometimes the ingates are choked.

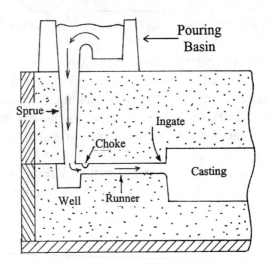

77

Gating systems are often described by ratios such as 1:4:4 or 1:3:3. This is the sprue-runner-gate area ratio. In a 1:4:4 system, the total runner area is 4 times the choke area (sprue base) and the total gate area is also 4 times the area of the choke. Some aerospace castings use a 1:8:4 ratio to allow impurities to separate out in the runner system.

Solidification starts at the thin sections of a casting and they draw molten metal from the thick sections. Thick sections of the casting or high shrinkage alloys need **risers** or reservoirs where the metal stays liquid long enough to feed the thicker casting as they freeze. As the metal solidifies and shrinks, it draws more liquid metal from the risers to fill in the shrinkage. The casting should solidify before the risers. A riser may be located on the top or the side of a casting and may be open or blind. An **open riser** is open at the top of the mold and a **blind riser** is buried in the sand.

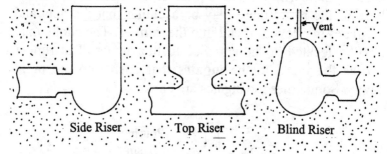

Side Riser   Top Riser   Blind Riser

Riser height and the neck dimensions are important because the riser must stay liquid longer than the casting. Riser distance or distance from the casting is important in side risers. Risers that are too close to the casting may overheat the sand and cause a hot spot or shrinkage defect in a casting.

Each of the parts of the gating system will be examined in detail. First, we will identify the key functions of the gating system.

1. Minimize the turbulence of the metal as it flows from the top of the mold into the mold cavity. Turbulent flow causes mold gasses, air, dross and or slag to be entrained in the molten metal and is a principle cause of defective castings. Aluminum and silicon bronze are particularly sensitive to turbulence and air aspiration.

2. The gating system should slow the metal stream down to allow the slag and dross to float out and stick to the tops of the runners.

3. Fill the mold cavity as fast as possible without creating turbulence. The mold should be filled quickly to prevent cold shuts or premature freezing before the mold is completely filled.

4. The gating system should promote directional solidification by delivering hot metal to the thickest section. The last metal to solidify should be in the risers, therefore it should be the last metal to enter during pouring. (There are situations where the risers are not the last fed and these are discussed later).

POURING BASIN:

When the total amount of metal poured exceeds 90 cubic inches, it is advantage to use a pouring basin. The best pouring basins are rectangular with a flat as opposed to rounded bottom. The basin should be large enough to "hit" when pouring the molten metal and should be deep enough to prevent a vortex from forming and drawing air into the sprue. Two and a half inches minimum depth works well for most smaller castings. The basin may be formed from green sand, however they are often made of baked sand. A basin is easily formed with a mold, as seen in the photo on the next page. Pouring cups are less desirable than basins but also used. They are easily formed in wooden or metal core boxes, as shown at the beginning of the chapter.

Pouring Basin Formed using Core Sand

SPRUES:

Sprues should be tapered with the small end being down.

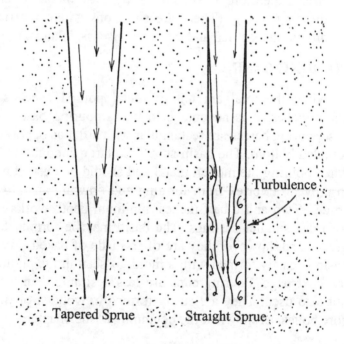

Turbulence

Tapered Sprue   Straight Sprue

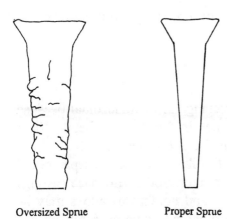

Oversized Sprue   Proper Sprue

The bottom of the sprue regulates the flow of metal into the gating system. If the sprue is too large, it does not fill properly resulting in sloshing of the metal, erosion of the sides and the formation of oxides.

At the bottom of the sprue there is a well or base. The bottom of the base should be flat and not round because round bottoms cause turbulence in the flow.

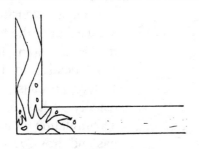

The well area should be five times the choke area and the depth should be two times the runner depth. Rectangular wells are better than circular wells.

CHOKE:

Chokes regulate the metal filling the mold. They may be formed in the runner or a core choke may be placed at the sprue base. It is never placed at the top of the sprue The diameter of the hole in the choke is determined by the required pouring rate.

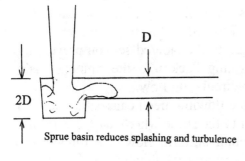

Sprue basin reduces splashing and turbulence

The calculations for choke design are given in an example later.

RUNNERS:

The best runners are rectangular or trapezoidal because this minimizes the turbulence in the flowing metal. When the ratios are large such as the 1:4:4, the metal flow slows down enough for much of the slag and dross to separate out and stick to the top of the runner. Some aerospace castings use a 1:8:4 ratio. Cast iron and nonferrous alloys may be poured with wide shallow runners to give more top surface to trap dross and slag. However, this also gives greater heat loss and frictional losses in the flow. Steel is usually poured with square runners to minimize the heat losses.

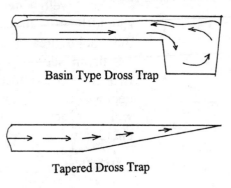

Basin Type Dross Trap

Tapered Dross Trap

Runner extensions are used after the last ingate to trap the first metal into the system because it usually has an accumulation of dirt, gas and dross. A few inches are usually enough and the end of the runner is vented so that gas pressure will not prevent the extension from properly filling. Tapered dross traps are preferred because the metal freezes in the tapered section preventing contaminants from washing back into the runner. Basin type dross traps cause a circulating flow.

The momentum of the flowing metal causes the metal to bypass the gates closest to the sprue when many ingates are used. In order to maintain uniform flow, the runners are stepped or reduced in size with each ingate.

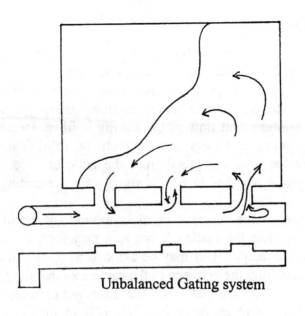

Unbalanced Gating system

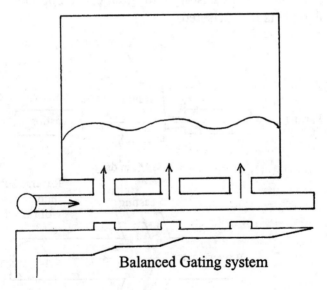

Balanced Gating system

Some foundries prefer to taper the runner rather than step it to reduce the chance of creating turbulence. Notice that the gates come off the top of the runner.

INGATES:

Ingates may be in the bottom, top or side of a casting. Each authority has reasons why one is better than the other, however, it really depends on the casting and what works best for the particular situation. Bottom gating gives a smooth, nonturbulent flow as the casting is filled. Because bottom gating is not as good for directional solidification, risers are fed with hot metal from the ladle or a second gating system is added to fill the risers after the casting is filled.

Because of the increase in mass around the point at which gates join the casting, a hot spot may form causing localized shrinkage. This may be avoided by gating into a riser and choking the entrance to the casting so that the gate freezes and draws metal from the riser rather than the casting. Gates that are more than half as thick or less than twice as thick as the casting will often cause hot spots and shrink defects if not risered.

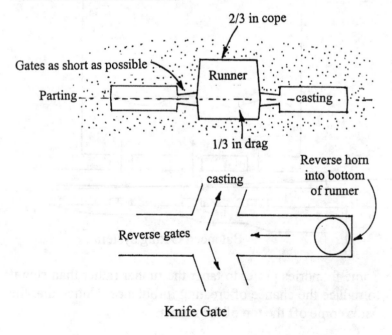

Knife Gate

84

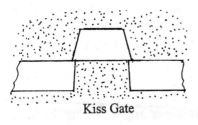

Kiss Gate

Wide flat gates work well and a small riser or shrink bob may be added to these gates. Knife and kiss gates can be used on small castings of long freezing rang alloys. Because knife and kiss gates are unable to supply feed metal they are not used on castings with heavier sections. These are not recommended for use on short freezing range alloys, or drossing alloys because of the turbulence they create. They are popular on smaller thin castings of long freezing range copper based alloys and some thin cast iron. Runners in these molds must be filled rapidly and any interruption in flow causes the gates to freeze. Molds are reverse gated or tilted to provide progressive filling of the mold cavities.

All gates must be clean and smooth to prevent turbulence and washing of loose sand into the casting. To avoid washing sand from a core, when possible, gate tangent to the core. Flat or plaque molds may be inclined 3 to 5 degrees to give smooth filling and not wash away the lettering (see copper practice).

Although they may produce a turbulent flow, pencil gates (about ¼ -inch diameter) are used to introduce a large amount of metal quickly into a thin casting. They are

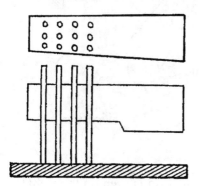

more often formed in baked sand rather than green sand.

Pencil gates may be formed in green sand by arranging the required number of gate sticks on the pattern. Sand is rammed around up to the required height in the cope. A basin pattern is drilled with holes that correspond to the

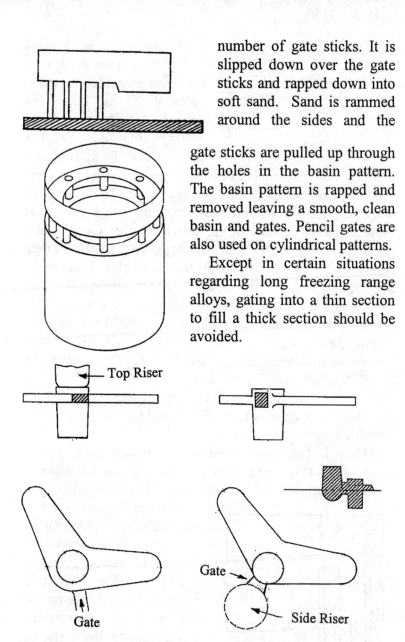

number of gate sticks. It is slipped down over the gate sticks and rapped down into soft sand. Sand is rammed around the sides and the

gate sticks are pulled up through the holes in the basin pattern. The basin pattern is rapped and removed leaving a smooth, clean basin and gates. Pencil gates are also used on cylindrical patterns.

Except in certain situations regarding long freezing range alloys, gating into a thin section to fill a thick section should be avoided.

Top Riser

Gate

Gate

Side Riser

Improved feeding of the heavy section is seen on the right.

FILTERS:

Using properly designed gating systems, impurities in the liquid metal will separate from the flow and remain in the gating system, leaving clean metal for the casting. When the gating design is less than optimal and when clean metal is essential, filters may be added to the gating system to remove slag, oxides and other impurities.

Filters are most effective when placed as close to the casting as possible. Many are used in traps or enlargements in the runner, or at the base of the sprue.

Filters may be as simple as a row of nails in the runner, or coarse steel wool. Commercial filters are available as small porous ceramic blocks or as woven ceramic sheets. A block type filter is inserted into an enlargement in the runner left by a print on the runner pattern. Sheet filters are available in several mesh sizes, are relatively inexpensive and easy to use. They do not require prints and may be inserted between the cope and drag at the gates or the sprue base. Sheet filters are easily trimmed to size using scissors. Currently the cost approximately $8.00 for a 12 x 12-inch sheet from which many filters may be cut. The 2-inch square filters seen below are supplied by Ametek.

Woven Ceramic Sheet Filters

DESIGN OF GATING SYSTEMS:

The majority of the castings poured in the small foundry will work well with sprues between ½-inch and ¾-inch diameter. For a few small non-critical castings, cutting the runners and gates by hand and a little trial and error will work. Larger or more difficult castings may require the design of a gating system. Frictional losses in the gating system and variation in the height of the ladle above the pouring basin among other things prevent absolute accuracy in the calculated flow rates. The formulas are far from perfect but they do give you a starting point from which the sprue dimensions may be adjusted .

The first step in designing a gating system is to determine the casting **weight** and the **minimum wall thickness**. From here you can determine the **optimal pouring rate**. The optimal pouring rate may be a range between a maximum and a minimum. The formulas for pouring time are derived from practical experience. They are most likely to produce good castings consistently.

The pouring time is governed by several factors that include the type of alloy, the fluidity of the metal, the pouring temperature and the weight of the casting.

After determining the pouring rate, **pouring time** is calculated by dividing the casting weight by the pouring rate. Finally the **area of the sprue base** is calculated.

The procedure for designing the gating system for iron, aluminum or bronze is similar. The main difference being the pouring time.

The sprue base area or choke area is calculated by:

$$A = W / (d \, t \, c \, \sqrt{2 \, g \, H})$$

A= area, W= Weight of casting,
d = density of the metal,
t = pouring time

c = Discharge coefficient for sprues	d= density of liquid metal lb/in^3	
.88 for round tapered sprue	Aluminum	.097
.47 for round straight sprue	Bronze	.266
.74 for square tapered sprue	Iron	.238

g= 386 in/sec^2 (acceleration due to gravity)

H = effective height of the sprue.
The area is converted to a diameter by:

$$D= \sqrt{4A / 3.14}$$

Example Aluminum Casting:

Design a sprue base for a 35-pound aluminum casting with ½ inch thick sections. With a basin on top of the cope and using a round tapered sprue, the effective sprue height is 7-inches.

1. From the aluminum graph 2, the pouring rate is 4 pounds per second.
2. Divide the weight of the casting by the pouring rate to find the time.

   35 pounds / 4 pounds per second = 8.75 seconds

3. Find the choke area: $A = W / (d\,t\,c\,\sqrt{2\,g\,H})$

   $A= 35 / (.097 \times 8.75 \times .88 \times \sqrt{(2 \times 386 \times 7)})$

   A= .637 inches^2

   Diameter = $\sqrt{4A / 3.14}$

   Diameter = $\sqrt{(4 \times .637) / 3.14}$ = .90 inches

POURING TIME FOR IRON:

Iron pouring *time* for iron is adjusted for the fluidity of the metal. By using the iron's composition and pouring temperature, the fluidity coefficient "K" is found on the graph and multiples the general pouring *time*. The pouring *rate* in pounds per second is found on the graphs and used to determine the general pouring time.

To calculate the pouring time for iron:

1) Find the composition factor
2) Find the pouring temperature
3) Find "K," the fluidity factor of the metal on the graph
4) Find the pouring rate on the general iron time graph.
5) Adjust the pouring time using K

To use the fluidity table, the carbon equivalent is calculated:

CE = %Carbon + ¼ Silicon % + ½ Phosphorous %

Estimate the pouring time for an iron casting that weighs 90 pounds and has a critical section thickness of ½ inch. It has a pouring temperature of 2600° F. The composition is 3% carbon, 2% Silicon, .1% Phosphorous.

1) Find the composition factor:

CE = %C + ¼ Si % ½ P %

CE = 3 + (2 ÷ 4) + (.1 ÷ 2)

CE = 3 + .5 +.05

CF = 3.55

2) The pouring temperature is given as 2600° F.

3) Find the Fluidity Coefficient K on graph 3

K = .68

4) Find the general pouring rate

The casting weight is given as 90 pounds.

Find the pouring rate on the graph 6, or substitute the values in the general iron rate formula were w = weight and T= thickness.

Pouring rate = $\sqrt{W}$ /(.95 + (T / .833))

Pouring rate = $\sqrt{90}$ /(.95 + .5 /.833)

Pouring rate = 6.1 pounds per second

Find the time:
90 pounds / 6.1 pounds per second = 14.75 seconds

Adjusted time = .68 x 14.75 seconds =10 seconds

POURING RATE FOR ALUMINUM CASTINGS:

The pouring time for aluminum casting is primarily related to the minimum thickness. The results are represented on the graph.

Thickness is less than .25 inches

Pouring rate in pounds per second = 1.48 $\sqrt{W}$

Thickness is .25 to .4375 inches:

Pouring rate in pounds per second is .125$\sqrt{W}$

Thickness is greater than .4375 inches:

Pouring rate in pounds per second is .698 $\sqrt{W}$

POURING RATE FOR BRASS AND BRONZE:

The pouring rate in pounds per second $= \sqrt{W} / .86 + 1.09T$

W = weight of the casting and T = the average thickness of the thinner sections in the casting

WELL or SPRUE BASE DIMENSIONS:

An enlargement at the base of the sprue called a well is used to slow the metal so that it enters the runners smoothly. Generally, the area of the to of the well is 5 times larger than the area of the base of the sprue. The well is twice as deep as the runner. Wells should be square or rectangular and must have flat bottoms.

Find the dimensions of the top of the well if the area of the sprue is .196-inches^2: 5 x .196 = .98-inches^2. The length of one side is $\sqrt{.98}$-inches^2 =.99 inches or about 1 inch.

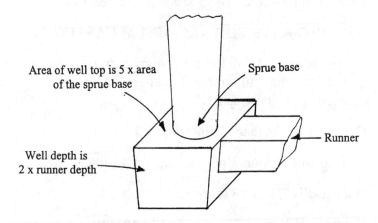

The remaining runners and gates are designed according to the desired ratio. The 1:4:4 ratio is commonly used in aluminum casting. If the sprue base has an area of .196-inches^2 (½ -inch diameter) then both the runners and gates would have a total area of 4 x .196 or .785-inches^2.

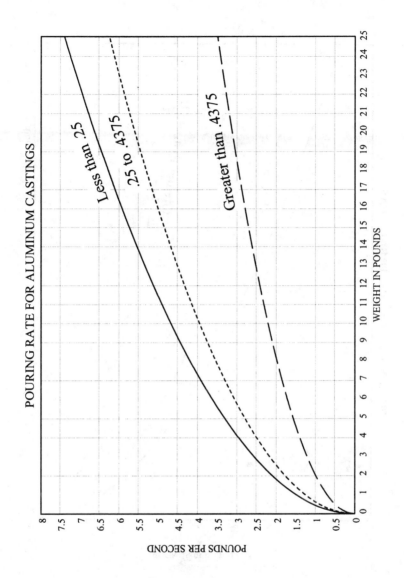

Graph1

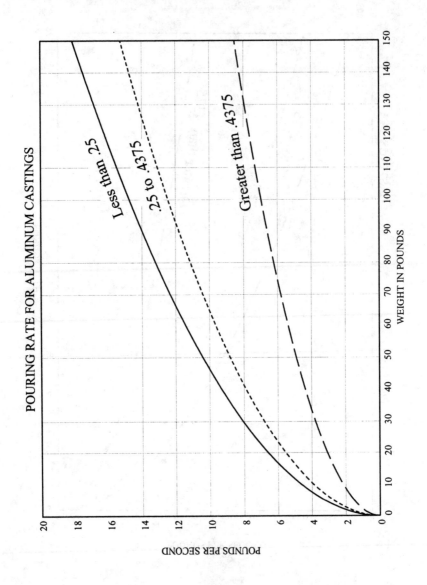

Graph 2

94

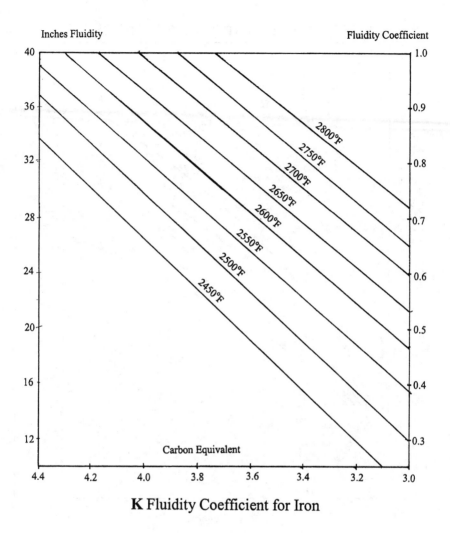

K Fluidity Coefficient for Iron

Graph 3

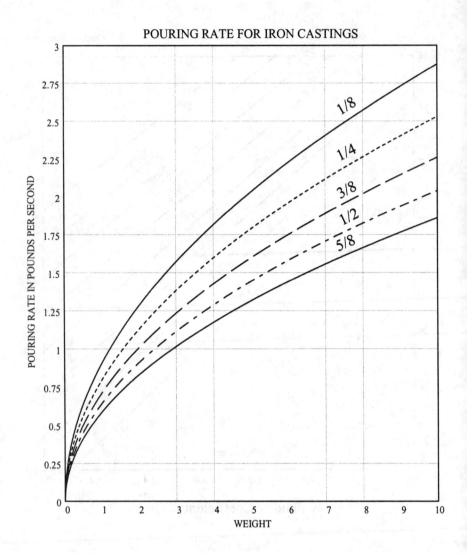

POURING RATE FOR IRON CASTINGS

WEIGHT

POURING RATE IN POUNDS PER SECOND

1/8
1/4
3/8
1/2
5/8

Graph 4

96

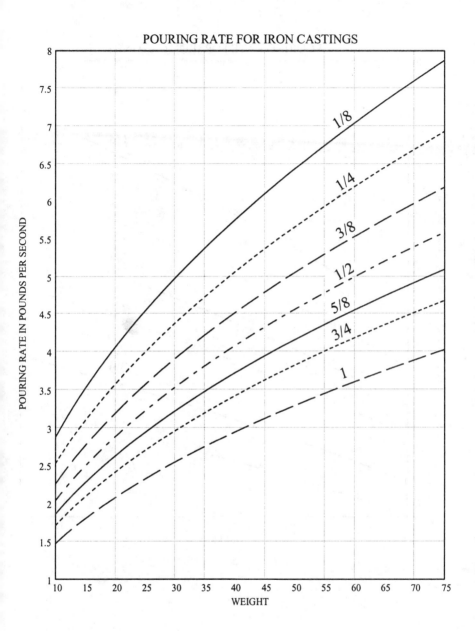

POURING RATE FOR IRON CASTINGS

Graph 5

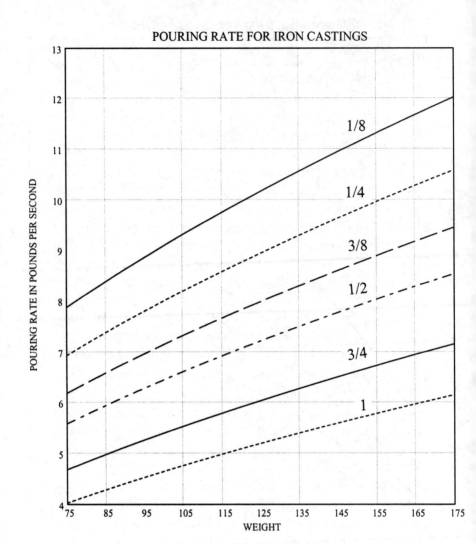

POURING RATE FOR IRON CASTINGS

Graph 6

98

VI. RISERS and FEEDING of CASTINGS:

Because long and short freezing range alloys solidify differently, no one set of specific guidelines can be given for the placement of all risers. General riser dimensions are given but should be modified to suit the particular job at hand. For the small foundryman, selection of proper risers is still a trial and error affair.

Guidelines that generally represent the short freezing range or skin forming alloys have been generated by years of experience in steel casting. In these alloys, shrinkage occurs as riser piping, gross shrinkage at hot spots and centerline shrinkage in uniform sections. For this situation, use hot risers gated directly from the runner when possible

Many aluminum alloys and long freezing range copper based alloys are not skin forming but freeze in a mushy or pasty state with dispersed micro-shrinkage. These alloys behave differently than short freezing range alloys. Heavy risering may not significantly improve the situation and may make it worse. Good feeding is better produced by steep temperature gradients towards the riser. This is accomplished by proper placement of chills and insulating boards. In some situations, micro-porosity is not a problem and the foundry seeks to distribute the porosity as widely as possible throughout the casting. This is accomplished by making it solidify as uniformly as possible. When section thickness is mixed, gating into thin sections, the placement of chills and dead risers on the heavy sections helps reduce the sink marks or depressions. The dead risers still must remain liquid longer than the casting so they should be insulated or topped with hot metal.

Riser Shape:

In order for risers to feed a casting, risers must remain liquid longer than the casting. Because solidification time is controlled by the rate of heat lost from the riser, those with smaller surface area relative to volume remain liquid longer. Spherical risers have the longest freezing time followed by cylindrical with a rounded head and cylindrical with a flat head. Rectangular risers have the shortest freezing time.

Riser Size and Shape	Surface Area in^2	Solidification Time, min.
6" sphere	100	7.2
4.25 dia. x 8" cylinder	120	4.7
Rectangle 3.625" x 3.625 x 8.625"	135	3.6

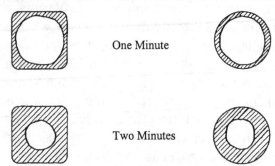

One Minute

Two Minutes

For an equivalent volume of metal, rectangular risers are very inefficient relative to other shapes. Spherical risers, while having the smallest surface area to volume ratio, are difficult to mold and not frequently used.

Heavy thick pads or shrink heads may be added to supply feed metal to sections of a casting as seen on the cylinder block casting above. These pads are machined off after casting to leave a shrink free surface

Riser size: Risers should be at least 1 to 2 inches higher than the section being fed. The diameter should be at least 20% greater that the thickness of the section being fed. Cylindrical risers are most commonly used and have an ideal height to diameter ratio for optimum performance. The height should be at least ½ the diameter and no more than 1 ½ times the diameter. Rounded bottoms have a lower surface area to volume ratio and therefore keep the riser liquid longer than flat bottoms. The same principle applies to the tops of blind risers.

Riser Necks: The feed metal must get from the riser to the casting through the neck. If the neck is too small, it freezes off and can not supply feed metal to the casting. If it is too large, it becomes difficult to remove. Some nonferrous alloys may require larger necks than specified below.

Necks for side risers: (see illustration next page) When possible, the neck should be circular to present the lowest surface area to volume ratio. All riser necks should be as short as possible. Circular riser necks should be no longer that ½ the riser diameter. Square and rectangular necks should be no longer than 1/3 the diameter of the riser. The diameter of the neck is (1.2 x length of neck + .1 x the diameter of the riser). For plate castings, the height of the neck is .6 to .8 the thickness of the casting.

Necks for top risers: The length of necks for top risers is the same as for side risers. The diameter of the neck for top risers is (length of neck +.2 x the diameter)

Blind Risers have many advantages. They can be made smaller and are more easily positioned. Blind risers are also more easily removed from a casting.

In skin forming alloys, the surface of the riser freezes over and a partial vacuum is created as the metal is drawn into the casting. In this situation, the skin of the riser is pierced and ventilated by using a carbon rod or dry sand core. Some prefer the sand core and claim the gas generated by the burning binder helps to pressurize the riser and further help feed the casting.

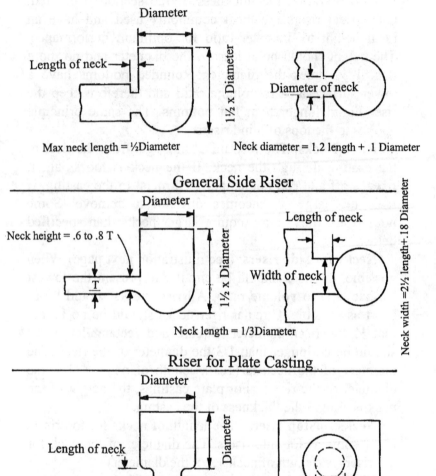

Max neck length = ½Diameter

Neck diameter = 1.2 length + .1 Diameter

General Side Riser

Neck length = 1/3Diameter

Riser for Plate Casting

Max Neck length= ½ Diameter

Neck diameter =Neck length +.2 Diameter

Top Riser

If properly ventilated, blind risers on skin forming alloys may feed a section higher that the top of the riser. However, if the skin on the casting is broken, the vacuum is lost and the riser will not properly feed.

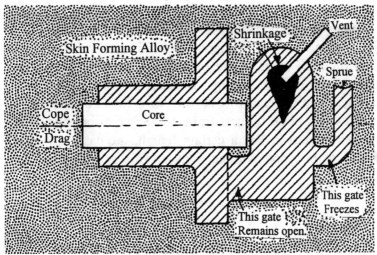

Skin Forming Alloy

Shrinkage

Vent

Sprue

Cope

Core

Drag

This gate Remains open.

This gate Freezes

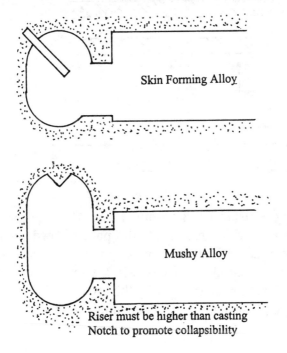

Skin Forming Alloy

Mushy Alloy

Riser must be higher than casting
Notch to promote collapsibility

Ventilating a blind riser in non-skin forming alloys provides no benefit because a vacuum is not created. In this situation the top the riser must be higher than the top of the casting. The shrinkage is fed by the weight of the metal forcing it into the casting. Notching the top of a blind riser helps it to collapse as it feeds.

The Feeding Distance of Risers is controlled by progressive and directional solidification. *Progressive solidification* starts at the mold wall and progress toward the center of the casting. Because of the greater heat extraction at the ends of a casting, *directional solidification* starts at the point farthest from the source of hot metal and moves toward it. For a certain length, the casting will be found to be good without a riser. This is called the *end effect*.

Progressive Solidification

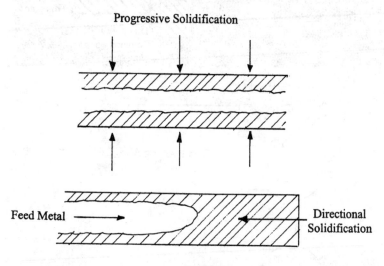

Feed Metal

Directional Solidification

End Effect

Feeding distance is limited by the progressive solidification, which closes off the path of feed metal. If progressive solidification moves faster than directional solidification, shrinkage voids will occur in the casting. Solidification may be sped up by the use of chills or slowed by applying insulating pad. Feeding distance may be improved by chilling the end farthest from the riser so that it solidifies before the progressive solidification cuts off the feed metal. Insulating the area closest to the riser, slowing the progressive solidification also increases the feeding distance. By tapering the wall thickness and using a combination of fins, chills, and insulating pads, the feeding of difficult sections may be accomplished.

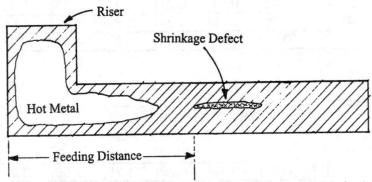

For aluminum castings the feeding distance (FD) is 2 times the section thickness (T).

$$FD_{Aluminum}=2T$$

Again, this may be modified as described above.

The feeding distance for steel castings is approximated by:

$$FD_{Steel}= 3.6\sqrt{T}$$

Plate with a chill:

$$FD_{Steel}= 11.6\sqrt{T}-3.2$$

Plate with end effect:

$$FD_{Steel}\ 11.6\sqrt{T} - 5.2$$

Plate with end effect and chill:

$$FD_{Steel}\ 11.6\sqrt{T} - 3.2$$

Useable results for copper castings may be approximated using the feeding distance formulas for steel castings.

Weight of Aluminum Risers

Height	Diameter in Inches					
	1	1.5	2	2.5	3	4
1	0.08	0.17	0.30	0.47	0.93	1.21
2	0.15	0.34	0.61	0.95	1.86	2.42
3	0.23	0.51	0.91	1.42	2.78	3.64
4	0.30	0.68	1.21	1.89	3.71	4.85
5	0.38	0.85	1.52	2.37	4.64	6.06

Weight of Iron Risers

Height	Diameter in inches					
	1	1.5	2	2.5	3	4
1	0.20	0.46	0.82	1.28	2.50	3.27
2	0.41	0.92	1.63	2.55	5.00	6.53
3	0.61	1.38	2.45	3.83	7.50	9.80
4	0.82	1.84	3.27	5.10	10.00	13.06
5	1.02	2.30	4.08	6.38	12.50	16.33

Weight of Brass & Bronze Risers

Height	Diameter in inches					
	1	1.5	2	2.5	3	4
1	0.25	0.55	0.98	1.54	3.02	3.94
2	0.49	1.11	1.97	3.08	6.03	7.88
3	0.74	1.66	2.95	4.62	9.05	11.82
4	0.98	2.22	3.94	6.15	12.06	15.76
5	1.23	2.77	4.92	7.69	15.08	19.69

Chills: A chill is something used as a mold wall at a point where we want quick solidification. Chills are used to speed up directional solidification and are especially effective with nonferrous alloys. A very hard surface forms when chills are used with cast iron. Because chills cause quick solidification before the feeding channel closes off, the use of chills increases the feeding distance of the riser.

A chill may be made of a block of metal, a rod or tube. Steel shot and chilling sands, which are discussed in Metal Casting Volume 1, may also be used as chills.

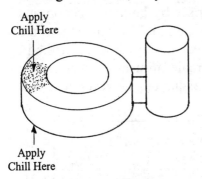

Apply Chill Here

Apply Chill Here

On a circular casting, a properly placed chill may also act like the end of a plate and promote directional solidification from a point furthest from the riser.

Chills must be clean and dry. It is good practice to sand blast them before use. Generally, chills should be the same thickness as the wall to be chilled. Thin chills may curl under the heat or fuse to the casting. Chills may be protected from adherence by painting them with shellac and dipping in fine sand while still wet. Additional layers may be built up as required.

Chills that are too large may cause cracking in a casting. Chills may be tapered to reduce cracking at the edges.

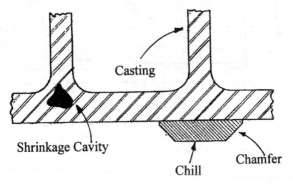

Casting

Shrinkage Cavity

Chamfer

Chill

HEAT LOSS FROM RISERS:

Heat flows from risers in three ways, (1) convection, which is transfer of heat from a surface to moving fluid or gas such as air. (2) Radiation. An incandescent lamp feels hot if you hold your hand near it. This is due to radiation. (3) Conduction refers to the heat that is transferred across an object. If you hold one end of a copper rod while the opposite end is heated, soon you will feel the temperature rise due to conduction.

Heat leaves the top of a riser by both convection and radiation and is transferred to the mold by conduction into the sand. Heat does not flow equally by all three methods. Temperature exerts a considerable influence on the rate at which heat is transferred by radiation; however it is less important in convection.

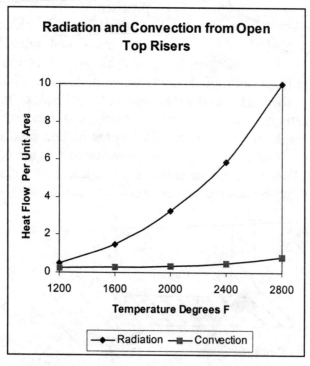

While convection may contribute significantly to the heat flow from an aluminum riser, it is almost negligible in an iron riser compared to radiation. Because radiation increases rapidly with temperature, it is more significant for the higher temperature metals. In copper alloys, convection amounts to approximately 10% of the total heat flow. Because any radiation shield will also significantly reduce convection, both terms are lumped together as "radiation" for the remainder of this discussion.

Heat flow into the mold is a complicated process that involves radiation from grain to grain, vaporization of water and flow of steam and gas between grains, and conduction through each grain. Again, all of this may be lumped as conduction. When metal is first poured into a mold, heat flow into the mold is very rapid and slows with time as the sand saturates with heat, however radiation is constant. More heat is lost through radiation on larger risers because conduction slows as the temperature of the sand rises.

Heat, like electricity or water, will flow along the path of least resistance. If one part of the path is restricted by insulation, the heat will flow through another path. When the sides of a riser are insulated, more heat will leave through the top; thus covering reduces radiation much more from insulated risers than from uninsulated risers.

	Open	Insulated	Covered	Insulated and Covered
Steel	5.0	7.5	13.4	43
Copper	8.2	15.1	14.0	45
Aluminum	12.3	31.1	14.3	45.6

Solidification Time in Minutes for 4 x 4 -inch Cylindrical Risers

Because radiation is slight at lower temperatures, applying insulation to an aluminum riser nearly triples the solidification time, while applying the same insulation on an iron riser (without a radiation shield) does not significantly change the solidification time. Applying a radiation shield to an iron riser more than doubles the solidification time. The significance of radiation on heat loss at higher temperatures is clear.

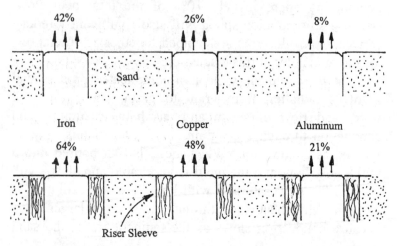

Per Cent of Total Heat Lost by Radiation from Open Top Risers

Riser shields may be made of gypsum, dry sand or charcoal. Because of the sulfur content, gypsum covers are not used on iron. Several excellent commercial riser toppings are also available and are discussed in the chapter making insulated riser sleeves.

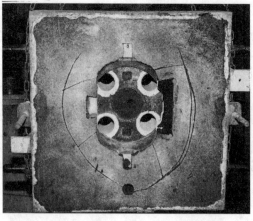

Insulating riser sleeves as seen from inside the cope.

MAKING INSULATING RISER SLEEVES:

Insulating riser sleeves for aluminum and bronze castings may be made from gypsum or plaster. Plaster sleeves do not work for iron castings because the sulfur in the plaster reacts with iron.

Howard Taylor and William Wick developed the original riser sleeves at the Naval Research Laboratory during World War II. There was a serious problem with casting quality, especially in pressure-tight castings. The problem was often linked to poor feeding. It was known that gypsum had good insulating properties relative to sand as seen below. Gypsum sleeves were made and a significant improvement in casting quality was noted when using the plaster sleeves.

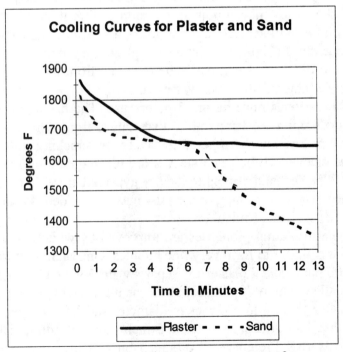

Manganese Bronze poured at 1880° F

111

The use of riser sleeves and topping compounds are common practice today. Modern sleeves have much greater insulating qualities than the original plaster sleeves and may be purchased from several companies.

Blind and open riser sleeves for aluminum and copper alloys are easily made from the wooden patterns shown at the end of the chapter. Because such a small amount of plaster is used in each riser, several patterns should be prepared. Begin by coating the patterns with a release agent such as petroleum jelly. Vigorous mixing of 40% liquid soap, 30% water and 30% oil also makes a good release agent.

Using a scale, mix the plaster using 1.6 pounds of water per pound of plaster. This consistency gives the least amount of cracking when drying the sleeves. When the plaster gets a creamy consistency and *starts* to set, pour it into the molds. If you pour it too soon, before it starts set, the plaster will separate out and make a hard bottom and a porous top. In about 15 minutes the plaster in the mold will set and the sleeves can easily be removed from the molds, provided the release agent was used.

The sleeves must be dried before use. The drying time depends upon the temperature that ranges from 400° F to 1200° F. The lower temperature causes less cracking of the dried sleeves. If the dried sleeves are too fragile, you may add 20% Portland cement to the dry gypsum before mixing. Other water absorbing materials may be added to the sleeves such as vermiculite.

The insulating properties are improved by whipping air into the plaster to form very fine bubbles. One such mixture is 100 parts dental plaster to 80 parts water by weight. Add a small amount of liquid soap. Mix the plaster with a rubber impeller about 4 to 6 inches in diameter and 1/8 inch thick inserted in a drill press. I use a 5-inch rubber-sanding disk from Harbor Freight tools mounted on a ½-inch diameter rod, 12-inches long. Hold the bucket of plaster so that air is whipped into the plaster and the volume increases about 50 to 70 percent. Pour the foam into the riser mold and allow it to set for 15 to 25 minutes. The sleeve is dried at 250° to 500° F. Higher strengths are found with the lower temperatures.

When molding, wooden riser plugs slightly larger than the riser sleeves are inserted in the proper locations. After the mold is rammed, the plugs are removed and the sleeves are inserted. Sand is packed around the loose sleeves.

If a mold is gated from the bottom and the risers are on the top, a second gating system may be used to fill the risers with hot metal after the casting is filled. The risers may be filled or topped off with hot metal after the mold has been filled and the risers are about 1/3 full. Hot metal is carefully poured into the tops of the risers. Sometimes the riser sleeves may rise above the surface of the mold by a few inches. In this case, fill the mold until the metal rises to the top of the sprue. Quickly place a piece of metal on top of the sprue to chill it off. The riser is then filled with hot metal from the ladle.

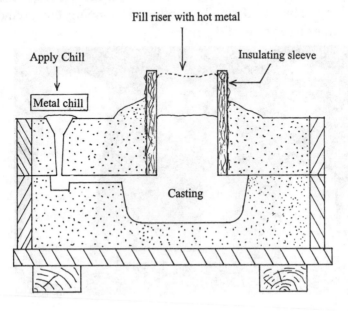

Plaster insulating pads may also be placed under risers. When used in combination with chills, favorable directional solidification may be achieved.

Perlite sleeves can be made in a similar fashion by using sodium silicate as a binder. The sleeves take less time to make; however they are not as effective as the plaster sleeves.

Exothermic (heat generating) sleeves and pads are made by mixing aluminum chips, an oxidizing agent such as sodium nitrate, a triggering agent such as magnesium chips, a binder and a sand filler (to control the rate of reaction.) The ingredients are mixed, molded and dried a low temperature in order to not ignite the sleeves. Exothermic sleeves are commercially available from many manufacturers.

Modern sleeves are made of superior insulating materials. They may be ordered in many diameters and can easily be cut to length with a knife. *Joymark* supplies a wide variety of insulating and exothermic sleeves (seen below). They also supply exothermic topping compounds to pour on the tops of the risers.

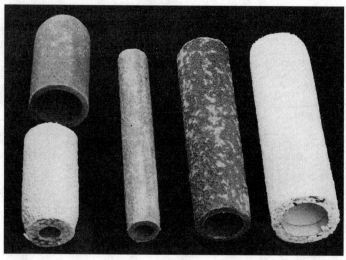

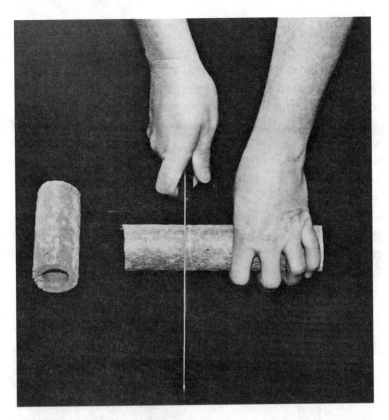

Cutting the Joymark Insulating Sleeves to Length.

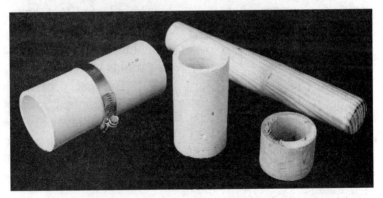

Porous Plaster Sleeve (center) made in a PVC Mold

Seen above is a porous plaster sleeve made in a section of PVC pipe that is split down one side with a hacksaw. The pipe is held together with a hose clamp. A wooden centering bushing locates the wooden inside-form in the center of the pipe. The inside form has a 1.5-degree taper in order to release the sleeve. The surfaces are lightly coated with petroleum jelly before pouring a plaster-cement mixture into the form. The joint between the centering bushing and the inside form also must be coated or the wood will swell and stick together. After the plaster sets, The sleeve is easily removed by loosing the hose clamp and spreading the sleeve by inserting a wood chisel into the saw cut. The inside form is removed and the sleeve is dried at 350-degrees.

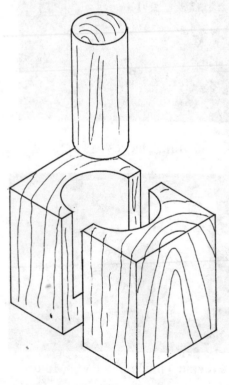

Wood Mold for Insulating Riser Sleeve

Riser sleeves are best cast in split molds. Because each sleeve requires such a small amount of plaster, several molds are needed to use even a pound of mixture. Wooden patterns may be adapted to a match plate allowing many aluminum riser-sleeve molds to be made.

Because of the sulfur content, plaster sleeves may only be used with aluminum and copper based alloys.

VII. PATTERN MAKING:

Pattern making could fill a book by itself and there are several good pattern-making books available. A pattern is the model of the part that is used to form the mold, in which the metal is cast.

When cooling from the molten state, all metals decrease in volume. When cold, castings will be smaller than their patterns. In order for the parts to be of the correct size, patterns are made a little larger to allow for shrinkage. This extra size is called the **shrinkage allowance**. Because the mold cavity is slightly enlarged by rapping and removing the pattern, the patterns used for small parts may be the actual size of the required casting. Larger parts require additional allowance for shrinkage of the cooling metal.

Aluminum, copper alloys and iron all have different shrinkage rates. Allowance for shrinkage is incorporated into a pattern maker's shrinkage ruler. These rulers are available for the shrinkage in common metals. You may make your own shrink ruler as outlined below, or you may calculate and add the percent change in dimension to your pattern dimensions.

Calculate the required change in length if aluminum shrinks approximately 5/32-inch per foot.

5/32 = .15625 There are 12 inches in a foot so divide by 12
.15625 / 12 = .013

To account for shrinkage in an aluminum casting, multiply all the dimensions by 1.013

If the required length of a finished casting is 9.5-inches, then the actual pattern length is 9.5-inches x 1.013 or 9.6235-inches.

117

If aluminum master patterns are cast for an aluminum part, double shrinkage allowance must be added to the original wood pattern.

Approximate Shrinkage of Metals in Inches per Foot:

Aluminum	5/32 –3/16
Brass	5/32
Cast Iron	1/8
Copper	3/16
Lead	5/16
Steel	1/4
Zinc	5/16

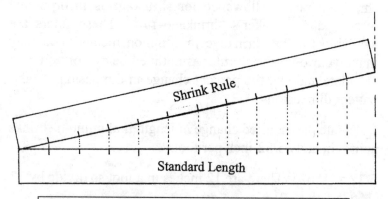

Graduating a Shrink Ruler Using a Standard Ruler

Machining allowance is the additional metal to allow for machining the casting. Depending upon the size of the casting, a surface may clean up with 1/16-inch machining allowance, however some castings may require ¼-inch or more. Machining allowance is always in addition to the shrinkage allowance. Clamping lugs may also be added to simplify mounting the casting for machining.

Avoid **sharp corners** on castings because they form stress risers, are difficult to mold and contribute to dirty castings. During pouring, sand washes off sharp corners corner as the hot metal flows over it.

Sharp corners are rounded with fillets. You can make fillets using Bondo and a pallette knife, available at craft stores. Traditionally, fillets were made with extruded bee's wax. Wax was heated and formed into a string. The string was then pressed into the sharp corners and smoothed with a

Fillet tool

fillet tool. The fillet tool is warmed in hot water to shape the wax more easily. A blow-drier also warms the wax for bending to shape. You may purchase fillet wax from a casting supply house or make a wax extruder. Bee's wax is found at craft stores.

Make the wax extruder from a 5-inch long pipe nipple. Chuck the nipple in a lathe and drill it to .875-inch diameter. Turn a piston from 1-inch diameter rod and drill a 3/8-inch hole through the center. Drill and tap a pipe cap 3/8-16 to accept threaded rod. Drill another pipe cap 1/8-inch or as needed for the extruder die. Weld a section of 3/8-inch diameter rod to form a T handle. Assemble as seen in the drawing. A free piston seems to have less reason to jam or bind than one tightly secured to the press screw.

Using a torch, gently warm the extruder starting from the tip and working down the length of the nipple. Allow a few minutes for the wax to heat throughout the pipe, then turn the screw. If the wax is well heated, it will come out in a spiral until it is enough long enough to pull straight. If it is too hot, it will spurt liquid wax from the end. The wax

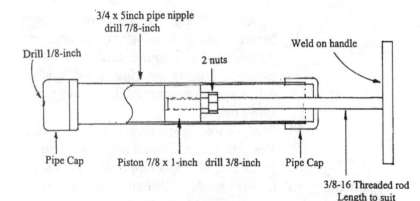

3/4 x 5inch pipe nipple
drill 7/8-inch

Drill 1/8-inch

2 nuts

Weld on handle

Pipe Cap

Piston 7/8 x 1-inch   drill 3/8-inch

Pipe Cap

3/8-16 Threaded rod
Length to suit

Wax Extruder

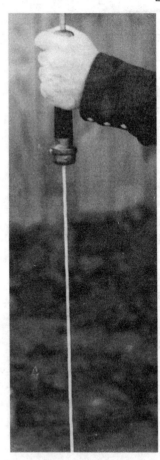

forms a jagged or rough string if it is too cold. It takes a little practice to get the proper heating technique.

Draft: In order to remove a pattern from the sand, the vertical surfaces must be tapered. Generally, on smaller parts, use 1½ degrees. Larger parts may use less draft however increased draft must be used when green sand projections extend from the cope. These have a tendency to break off and drop out when they are designed with sharp corners and little draft. Five degrees should be used for such projections.

Parting line: Flasks separate into the cope and drag at the parting line. The line that divides the pattern into the cope and drag sections is called the parting line.

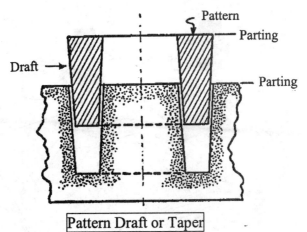

Pattern

Parting

Draft →

Parting

Pattern Draft or Taper

A pattern may have a flat surface or it may be split at the parting line.

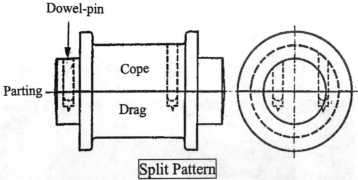

Dowel-pin

Parting

Cope

Drag

Split Pattern

Some patterns do not separate along a straight line and are said to have an irregular parting. A follow board is used for this type of pattern or if only a single casting is needed, coping out may form the parting line.

A quick way to make follow boards: Make a small box deep enough to reach the parting line of the pattern. To release well from the cope, cut the sides of the box with 45° angles. Coat the pattern with a release agent such as petroleum jelly. Using small wedges, level it in the box and fill the box to the parting line with plaster or Bondo. Smooth the plaster or Bondo as needed to form the parting.

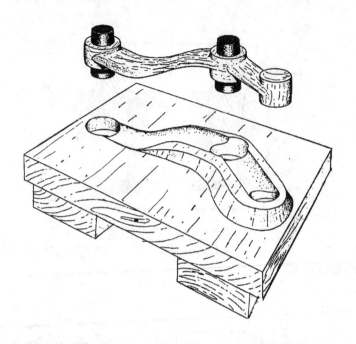

Follow Boards are used for irregular parting lines

This box is only as deep as the parting line and will be filled with plaster or casting resin to form a follow board.

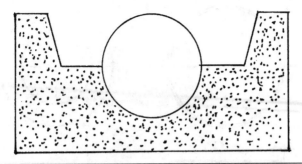

When coping out, ram the pattern in the drag and then cut away the sand to the parting line

Split Patterns are held in alignment to each other by dowel-pins. These pins may be made of wood or metal and should fit snugly but not tightly. The pins should be rounded on the ends and project about 3/32 of an inch. The pins may be applied by assembling the pattern and drilling holes completely through the cope and into the drag section. The pins may also be located by driving a small brad into one of the parting surfaces. Snip the brad off leaving about .1 inch extending above the surface. Assemble and clamp the pattern to leave an impression in the mating surface. To prevent improper assembly, never locate dowels symmetrically on a pattern. This makes it impossible to assemble the pattern incorrectly.

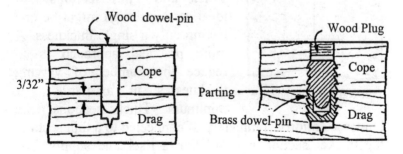

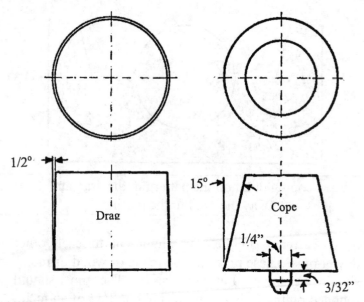

Drag

Cope

1/2°

15°

1/4"

3/32"

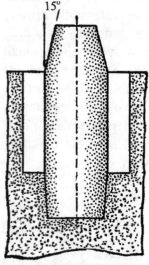

15°

Vertical Baked Sand Core

Core prints: Vertical core prints are tapered differently for the cope and drag. Drag prints have a slight taper, ½ to 1 degree, in order to hold the core straight while the cope has a large taper to easily guide it into its seat. The cope print is often equipped with a dowel pin to make it removable.

Thick and Thin sections: An ideal casting design might be one having only a single thickness. A more practical approach is to reduce the number of varying thicknesses in a part to a minimum. Thick sections take longer to solidify completely. If connected to a thin section, the thicker section will act like a riser and supply feed metal to the thin section causing the thick section to form shrinkage cavities. Also if a thinner section is attached to a thicker section, the thinner section will solidly before the thicker section and be in the process of cooling and contracting, while the heavy section is still solidifying. This results in different rates of contraction and forms stress in the casting. Cracking may occur at the weakest part of the casting, which will be the semi-solid thick section.

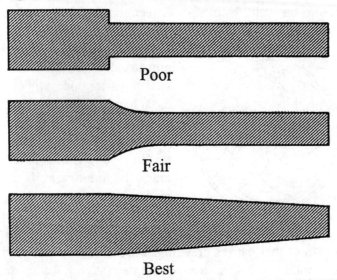

Poor

Fair

Best

In order to prevent stress concentrations and to promote directional solidification, joints between different section thicknesses should be tapered.

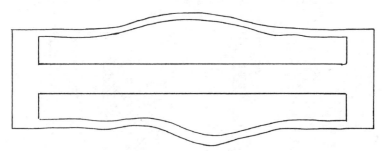

A casting where light and heavy sections are combined may shrink to produce distortion as seen above. The light bars solidify from the middle to the ends drawing feed metal from the heavier sections, thus reducing shrinkage. After the outside bars solidified, the heavier ends and middle sections solidify and contract exerting a high compressive force on the lighter bars causing them to buckle.

Because of the principle described above, long runners should be split into shorter sections and poured from multiple points to reduce distortion in long castings.

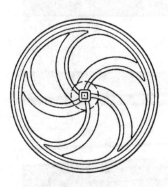

The rim and hub of a wheel are usually heavier than the spokes of an open wheel such as a hand wheel or flywheel. Because they cool at different rates, considerable stress is set up in the casting that may crack the hub or rim. If curved or S shaped spokes are used they will tend to be pulled toward straightness reducing the stress that causes cracking. Also using an odd number of spokes reduces the direct stress of two spokes contracting opposite of each other.

High Stress Concentration

Proper Fillets Prevent Stress concentration and Promotes Shrink-Free solidification

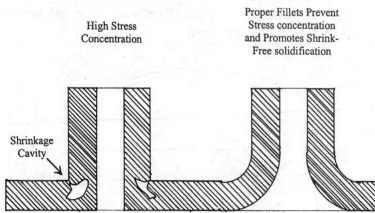

Shrinkage Cavity

Cracking occurs at the thicker joints

Too strengthen a pattern, or to prevent distortion during casting or heat treatment, tie bars or stop offs may be used. If the tie bar is filled with sand in the mold, it is called a stop off and does not become a part of the casting. Tie bars are used on U shaped castings and may also be used for gating. The tie bar is removed after casting or heat treatment.

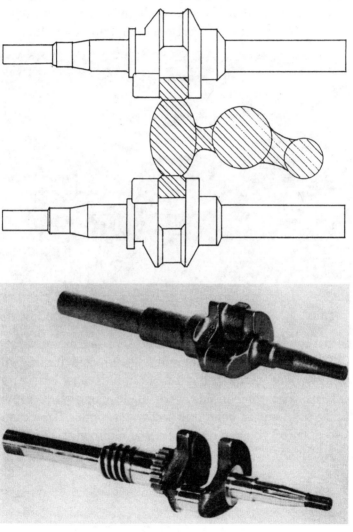

Pattern Making Tools: Common woodworking or traditional cabinet making tools are used for pattern making and include but are not limited to: Band saw, Table saw, Jointer and Plainer, Wood chisels, Drill press, Disk, Drum and Belt sanders, Lathe, various C and Bar clamps. An air driven brad nailer is very useful and inexpensive from tool suppliers such as Harbor Freight. *One indispensable tool is the dial caliper* that is used for both drawing and pattern making. Pallette knives and Bondo are good for fillets.

This detailed solid aluminum green-sand casting of a handgun was made using rubber molds and cast plastic patterns. The sand molds were made using #180 Olivine facing sand with a bentonite and water binder. This casting is so detailed you can read the serial number on the back of the handgun.

MAKING RUBBER MOLDS:

When making several identical patterns for a match plate, or when making a pattern from an original part, rubber molds can be used. Duplicate patterns are made by making a rubber mold from the original pattern. Plaster or plastic casting resin is then poured into the rubber mold to make identical copies of the original pattern

Properties of mold rubber: Depending upon the mixture, cured rubbers and plastics have different hardness or stiffness. These are measured on a scale called the "Shore A" or "Shore D" hardness. Rubbers are found on the A scale and plastics are found on the D scale.

RUBBER	SHORE A		VERY HARD
00 20 30 40 50 60 70 80 90 95			
VERY SOFT		45 55 65 75 85	
		SHORE D	PLASTIC

Viscosity or resistance to flow is measured in centipoise. Water has a viscosity of 1 centipoise, motor oil = 250 and molasses = 2500. Uncured liquid rubber may have a low viscosity and pour like water or they may be thick like molasses. Bubbles quickly rise to the surface of low viscosity rubber but high viscosity rubber retains entrapped air and must be vacuum degassed.

There are four types of rubber used for molding. Each has certain properties that make it useful in certain situations. Generally, sand casting foundries are interested in silicone and polyurethane molding rubbers. Low temperature alloys may also be cast directly into silicon rubber molds.

Latex is a natural rubber from rubber trees. Because latex is a one component system that does not have to be mixed, it is ready to use right out of the container. Latex is abrasion resistant and very elastic. It retains its shape after being repeatedly rolled up. Latex is almost always brushed on and may require as many as 20 coats with 4 hours drying time between coats. Latex usually shrinks about 15%. Latex may be used to cast wax and plaster patterns. They are not suitable for casting plastic resins.

Polysulfide rubbers are a two component system consisting of a base and a curing agent. Polysulfide mold rubber is a favorite of bronze foundries who make wax patterns. Polysulfide rubbers are very soft and elastic and are good for making molds with extreme undercuts and very fine detail. Polysulfide rubbers are mixed by weight and require a triple beam balance or similarly accurate scale. Wax patterns for the lost wax process and plaster patterns are made with polysulfide rubbers. They are not suitable for casting resins.

Polyurethane rubbers are relatively low cost, two part systems. They are easy to use and may be poured, brushed or sprayed on. Generally, polyurethanes are mixed by volume and do not require a scale. They have low viscosity and do not trap air bubbles. Flexible urethanes are available in a wide range of hardness or stiffness.

Polyurethane rubbers are moisture sensitive and may bubble if mixed when the humidity is high. They have a limited shelf life after opening unless they are covered with a dry gas.

Polyurethane rubber has extremely poor release properties and will stick to almost anything. The original pattern must be sealed and coated with a release agent. The release agent is sprayed on and then brushed into detail and undercuts to be sure that the pattern is completely coated. Do not soak the pattern with release agent as this causes pinholes on the surface of the finished mold.

Silicone rubbers are two component systems that must be accurately weighed on a triple beam balance or equivalent system. They are cured by either platinum or a tin catalyst. The tin molds shrink slightly and have a shorter life. Silicone may be poured, brushed or sprayed on to the original pattern and has the best release properties of all the available molding rubbers. Silicones are excellent for casting plastic resins and do not require a release agent. Because of its high heat resistance, low temperature alloys may be cast directly into silicon molds.

Silicone rubbers are more expensive than the other rubbers. They are sensitive to sulfur. Because some clays contain sulfur, they may not cure if exposed to unsealed clay patterns. Krylon clear acrylic spray paint makes a suitable sealer for patterns and clay. Silicone rubbers are very thick and must be vacuum-degassed to remove trapped air.

Making A Silicone Rubber Mold: Silicones will reproduce the finest detail right down to a fingerprint. They may be used to cast both plaster and wax patterns. Because plastics are easily cast in silicone rubbers, they have the best release properties of all the available rubbers and require vacuum degassing, silicone is used in the following demonstration. SMOOTH-ON MOLD MAX 30, a silicone rubber with a Shore A hardness of 30, is selected. It has a pot life, or working time of about 45 minutes.

1. Build a temporary flask. This can be made from a cake pan, box or from Masonite strips as seen on the next page. Wood blocks are glued and nailed to each end of the Masonite strip to form an adjustable flask. Seal the Masonite with acrylic spray paint such as Krylon to improve the release properties of the flask.

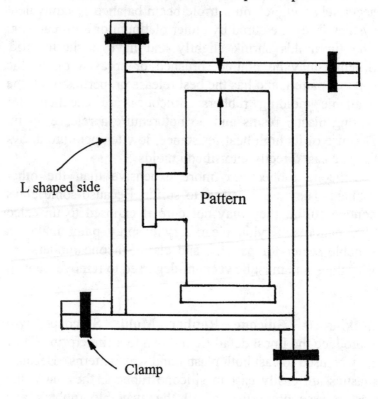

Maintain at least 1/2 inch clearance between pattern and sides and over top of the pattern

L shaped side

Pattern

Clamp

2. Cover the work area with paper. Heavy brown builder's paper or newspaper works well. Dress in an old shirt, with long sleeves and wear rubber gloves. The thin latex gloves are fine. Because silicon is very thick and somewhat stringy, the first time you mix you will probably make somewhat of a mess. Uncured rubber may be cleaned up with acetone.

3. Center the pattern on a piece of Masonite. The pattern may be permanently attached or secured with shellac. Be sure that there is no gap between the pattern and the molding board or the liquid rubber will seep under the pattern.

4. Clamp the sides of the flask so that there is at least ½ inch clearance between the pattern and the sides.

5. Seal the seams of the flask with modeling clay. Be sure to maintain the ½ inch clearance for the sides. Spray the seams and assembled pattern box with Krylon crystal clear acrylic.

6. Mix the mold rubber. Components for silicone rubber must be accurately weighed on a triple beam balance or an equivalent gram scale. *Improper weighing and poor mixing of components cause the mold rubber not to cure.* Because silicon rubber is very thick and difficult to mix, it is easier to divide the total amount of rubber needed into several smaller amounts. Measure approximately 350 grams of rubber into several large

plastic cups and then add the curing agent. Mix by scraping the sides and bottom of the cup and drawing it into the center of the rubber. Mix for approximately 2 to 3 minutes. Some of the unmixed rubber will stick to the bottom of each cup and will be mixed in the degassing container. When all the cups are mixed, pour them into a degassing container scraping the bottom of each cup. Mix for another 3 minutes. The degassing container must be 3 to 4 times larger than the total amount of rubber.

7. Apply a vacuum to degas the rubber. The rubber will swell to 3 or 4 times its original volume as the bubbles rise to the top. Maintain a vacuum of 28 to 29 inches Hg for at least 3 minutes. When the rubber falls back to its original volume, degassing is complete.

8. Be sure that the mold is level. Pour the mold rubber into a low point in the mold box and let the rubber slowly rise up over the pattern. This displaces the air from the pattern surface. Do not pour directly onto the pattern because this traps bubbles at the pattern surface.

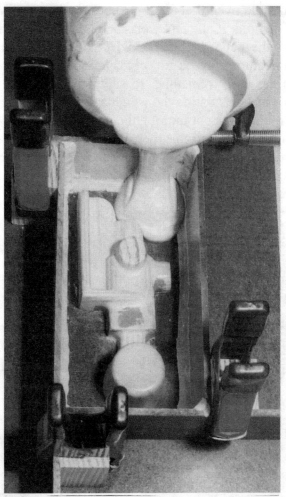

Small bubbles may rise to the top surface when the mold is filled. You may dissipate these bubbles by lightly heating the surface with a blow dryer.

9. Allow the mold to cure 16 hours at 77° F. Lower temperatures require longer cure times and below 65°, the mold may not cure at all. If the mold is placed in an oven and held at 150°F, it should cure within 4 hours. Because molds may leak, it is wise NOT to use the household oven. Let the mold cool to room temperature and demold. If all went well, the mold easily pulls away from the pattern. If the mold did not cure, the room temperature may be too cold, an accurate gram scale was not used (remember to subtract the weight of the cup when calculating the weights of the components) or the rubber was not thoroughly mixed. Sulfur clays may also prevent curing; therefore the mold is always sprayed with Krylon crystal clear acrylic to seal the clay.

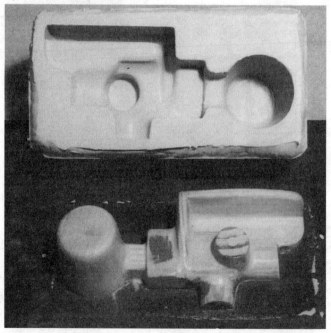

Completed Rubber Mold and Original Pattern

136

10. Optional post cure. To dry the rubber of any residual water or alcohol left from the curing process, the cured molds may be place in an oven at 125° F for 4 hours.

Making a vacuum chamber: A vacuum chamber may be assembled for about $30 or less. A section of pipe forms the walls of the chamber and two pieces of ¼-inch thick plexiglass forms the top and bottom. A vacuum gauge (Grainger part# 6X830), a few hose barbs and some ¼-inch ID vinyl tubing complete the chamber. A venturi type vacuum pump may be built or purchased from *Harbor Freight Tools* for $9.99. The vacuum fitting should be removed from the Harbor Freight pump and replaced with a hose barb.

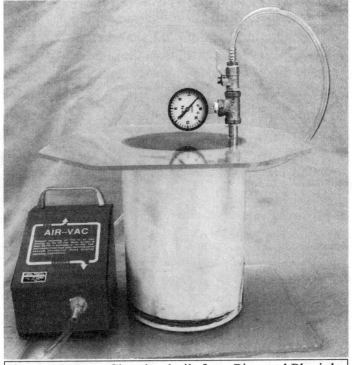

Simple Vacuum Chamber built from Pipe and Plexiglas

Chuck the pipe in the lathe and make a facing cut on each end. Large pipe must be run at a low rpm. Several light cuts may be required to smooth a rough sawn pipe. If the ends are properly cut, the vacuum chamber may be used without gaskets. Simply apply the vacuum and lightly press on the top to start the seal. The chamber should reach 28-inches of vacuum in 15 to 25 seconds using the small Harbor Freight pump.

Liquid Plastic: There are 3 main types of plastic resins used. *Urethane* resins are general-purpose resins and are available in a range of hardness. *Polyester* resins are found in Bondo, they shrink over time. *Epoxy* resins are used for industrial parts.

SMOOTH-ON liquid plastic SC-320 is a general purpose urethane resin and is used in the following demonstration. It has a Shore hardness of 75D a working time of 2 to 3 minutes and may be stripped from the mold in 10 minutes.
1. Cover the work area with paper, wear long sleeves and rubber gloves.
2. Level the mold.

Pouring liquid plastic into the lowest part of the mold

138

3. Shake both part A and B of the mixture before dispensing.
4. Using equal amounts by volume, mix the two components for 90 seconds, scraping the sides and bottom of the container. Remember the working time is only about 2 to 3 minutes.
5. When thoroughly mixed, pour the liquid plastic directly into the lowest part of the mold without degassing. Allow the plastic to rise slowly up into the mold as done when using liquid rubber. Be careful not to over-fill the mold causing the back to rise above the parting line.
6. In approximately 5 to 7 minutes the resin will change color and solidify. In 10 minutes the plastic part may be removed from the mold.

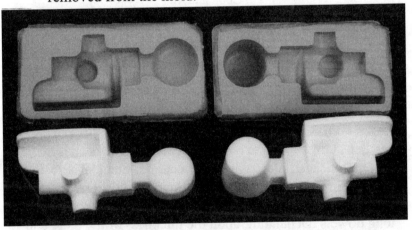

Completed cope and drag halves of the air valve

Urethanes are moisture sensitive and have a short shelf life once the container is opened. The shelf life may be extended by covering the unused resin with a dry gas such as nitrogen or argon from your welder.

MOUNTED and MATCH-PLATE PATTERNS:

When making more than a few of a particular casting, it may be worthwhile to mount the pattern on a plate with the gating and risers. The cope and drag portion of the pattern are mounted on the opposite sides of a wood or metal plate called a match-plate. The patterns may also be mounted on separate cope and drag plates, molded individually and assembled into a completed mold. Match-plates considerably speed the molding process and produce cleaner castings than loose patterns and hand cut gating. While match-plate molding is generally used with smaller patterns, large molds can also be produced this way.

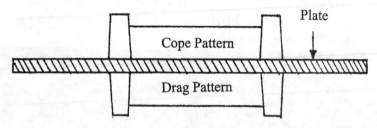

The best hand-held match-plates are made of ¼-inch aluminum plate. Furniture grade ¾-inch plywood also works well. There are several pattern mounting schemes and depending upon the required accuracy, one is chosen.

The quickest, although least accurate, method is to lay the patterns out on the plate, then mark and drill over sized dowel holes in the plate. Pin and screw the pattern with long dowels. Depending upon the thickness of the plate and accuracy of the original dowel holes, the patterns may or may not be matched. If the drill runs off, the thicker the plate, the greater the mismatch will be.

The plaster transfer is the most accurate method of pattern mounting. You must start with a flask that has accurate pin bushings. When properly machined, the aluminum flask found in Metal Casting Volume 1 makes an excellent pattern-mounting flask. The plate should also have accurate pin bushings. These bushings may be made by carefully step drilling a 1-inch diameter hole in a section of ¼-inch aluminum plate. These sections are split in half and mounted to the match plate forming adjustable pin guides.

Saw plate in half

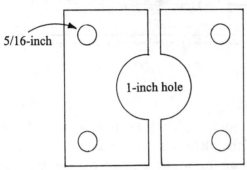

5/16-inch

1-inch hole

1/4-inch aluminum plate

To mount patterns using the plaster transfer method, start with regular patterns that are properly split and doweled.

1. Adjust the plate pin bushings for a proper fit and assemble the flask with the plate between the cope and drag. Using vise grips or suitable clamps, clamp the assembly together

2. Set the drag patterns in place. When you are satisfied with the layout, they may be attached using a little wax or shellac. The patterns *must sit flat* against the plate or they will be misaligned in the final assembly. Coat the patterns, plate and the inside of the flask with a release agent such as petroleum jelly dissolved in mineral spirits.

3. Mix plaster into water. A less stiff mixture both flows better around the patterns and is easier to chip out later.

4. Fill the flask with 1½ to 2½ inches of plaster. This should be enough to hold the patterns in place. As the plaster starts to set up, with a wet finger, clear away the

141

areas on the pattern where you intend to drill and insert dowels. You do not have to make a large hole, just enough to properly see and fit a drill bit.

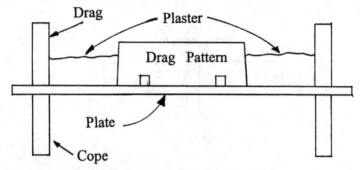

5. When the plaster sets, roll the flask over and remove the plate. If the plate is stuck, heating the back of the plate or heating the whole assembly in an oven at low heat may soften the wax or shellac.

6. Clean off the patterns and the parting face. Set and clamp the cope in position. Place the cope halves of the patterns in place and coat with a release agent.

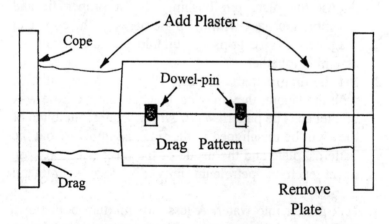

7. Mix and pour plaster around the cope pattern.

8. When the plaster has set, open the flask and remove the pattern dowels. Set the plate into position and close the flask.

142

9. Turn over the drag side, drill the new dowel holes, glue and drive the dowels into place.

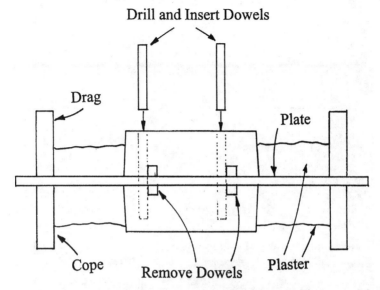

10. Using a hammer and large screwdriver, chip away the plaster starting at the edge around the flask. When the plaster is cleared from around the flask, remove the flask from the plate. Continue carefully chipping the remaining plaster from around the patterns.

Match-plate molding requires straight flask pins, a good vibrator and straight draws. Broken molds are usually caused by poor ramming, poor draws (not straight) or dry sand. It may or may not be worth attempting a mold repair. Sometimes the broken spot may be dusted with wheat flour to glue it into position. The mold is then closed and rammed over the bad spot. The mold is opened and the bad spot is pinned with a few nails.

A few example match-plates and their castings are seen on the next few pages. Large patterns may also be plate mounted.

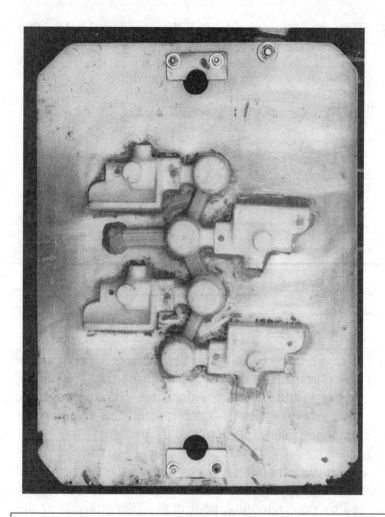

The plastic patterns from the rubber mold chapter were mounted on this plate using the plaster transfer method.

Completed Castings

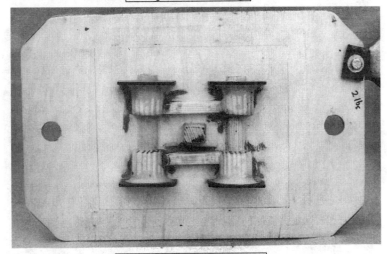

Flask hardware patterns

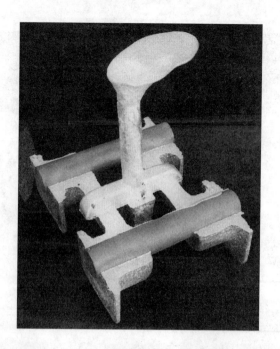

Above: Completed flask hardware casting with chills in place. Note that the wooden match-plate is labeled with the weight of the competed casting. Below: Match-plate for a Cincinnati Horizontal mill.

146

Top: Match-plate removed, cores set in place. Bottom: Completed casting

VIII. FOUNDRY PROJECTS:

MAKE A STURDY FLASK LATCH:

Finding sturdy, yet affordable, flask hardware is a problem for the small foundry enthusiast. Most of the available latches are poorly constructed and quickly break or loosen up after a few molds. The latch detailed below was adapted from the patent drawings in Volume I. It is very strong and cannot be broken by either hand ramming or a jolt machine. It is very inexpensive to build and requires almost no machining. It may be built using the scraps from the other projects in the series. Enough latches for two flasks may be easily assembled in an afternoon.

A snap flask must be constrained in at least two directions, x and y, if it is to be properly held. The latch consists of 3 pieces; a hinge and locking bar, an x restraint and a y restraint. Although the latch appears to be complete without the y restraint, it will pull open diagonally making an unsuitable mold if it is not included.

In order to get a satisfying "snap," the locking bar opens in a circular motion rather than a straight motion. Varying the radius of the locking bar controls the tightness of the snap. If the arm is too short, the latch becomes almost impossible to open. If too long, the snap is loose.

Because all flasks are not identical, general dimensions are given and the final fitting is done on an individual basis. After your first latch is fitted, you will know your tolerances and your remaining latches will fit quickly.

Making the Latch Plates: Cut 2 plates from either 16 or 18 gauge steel. The radius is rough sawn and cleaned up using a bench grinder and file. Mark and drill the holes for the mounting screws. I prefer to layout and mark the holes for several hinges on a strip of steel. Drilling them all a once while they are still connected in the strip. Cut the individual hinges from a strip *after* drilling.

Using one plate, mark a line ½-inch from the end, clamp the plate in a vise and make the right angle bend using a hammer.

Lightly paint (yellow is good) a 5-inch length of 3/4inch by 1/8-inch thick angle. Scribe a line .635-inch up from the bottom. Clamp the angle in a vise and grind to the line. Cut the angle into a 3-inch section and a 1.5- inch section (there will be a little left over). Set the 3-inch angle on the unbent plate and weld into position, as seen in the drawing. Drill the 2 mounting holes in the 1½ -inch section.

Making the hinge: Drill the inside diameter of a section of ¼-inch pipe to 3/8-inch. This is easily done in a lathe, but may also be done using a hand drill and vise. Note that the inside diameter of a ¼-inch pipe is .365 making this a very minor drilling operation. Test the fit by sliding a 3/8-inch diameter steel rod through the drilled pipe. Be sure to round the ends of the rod and remove any burrs on the pipe. When you are satisfied that the rod easily slides through the pipe without excessive binding, cut it into a 1½ -inch long piece and two ¾ -inch pieces as seen in the drawing. Cut an additional 3-inch section from ¼-inch pipe or ½-inch diameter steel rod to form a locking bar. Note that this piece does not have to be drilled. Assemble the hinge, set it flat on a welding bench or steel sheet and weld it to the bent plate.

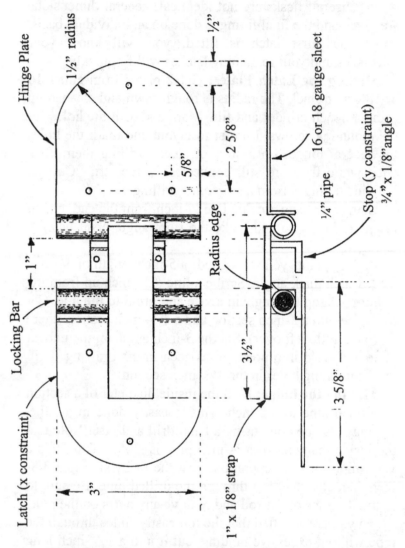

Cut a 3½-inch strip of 1-inch by 1/8-inch steel strap. Set the hinge on to the welding table and insert a shim, a squarely cut section of wood will do, between the locking bar and hinge. My shim is 1.05 inches wide by .4-inches

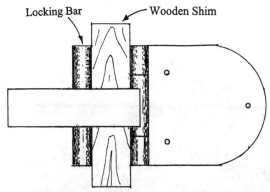

Locking Bar   Wooden Shim

high. This shim keeps the bar square to the hinge while welding. Weld it into position. Check to see that the hinge operates freely. Sometimes a little weld spatter will make it bind. Pick out or grind away the spatter to free the hinge. When you are satisfied with the operation of the hinge, give the completed latch a coat of paint. Wal-Mart sells spray paint for about 98 cents per can.

Assemble and fit the latch to your flask: Close your flask and clamp it shut using a bar clamp. The clamp should be tight enough to take up any slack in the flask but not so tight as to distort it. Place the hinge into position, drill and install the screws. Using a single flat head wood screw in the rear position pulls the hinge tight against the end of the flask. Set the plate with the welded angle into place and pull it tight against the locking bar. Screw the plate into position again using the flat head screw in the rear position. Insert the remaining screws, round head will do. Release the clamp and check the "snap." Grind or file the angle to get a satisfying snap without binding. When you are satisfied with the snap, install the y constraint as close the hinge as possible. You may have to shim or grind the y constraint to get the proper fit.

When properly assembled, the latch will close with a satisfying snap, will not pop open with hard ramming yet opens easily when pulled. You will be very pleased with your latch.

MAKE A KNEE OPERATED AIR VALVE:

A pattern vibrator is easily operated using a knee valve, freeing both hands to get the straight draw required when removing a plate mounted pattern from a mold. Knee valves are both difficult to find and very expensive to purchase. Fortunately they are neither expensive nor difficult to make. The valve project is introduced here because the air valve is used in a jolt squeezer project that appears later in the *Small Foundry Series*.

The valve is designed to use scrap materials left over from the other flask and furnace projects from *Metal Casting volumes 1 and 2*. It requires a ¼ 20 tap and a 1/8 and / or a ¼-inch pipe tap and a 1-inch diameter reamer. The reamer may be purchased or a D type reamer may be made from a section of 1-inch diameter drill rod.

The only critical part of the project is the placement of the inlet and outlet ports. The outside of the valve may be arranged to suit your mounting requirements. Both side and top mounting are illustrated.

Making the Pattern: Make a split pattern from two 10-inch strips of yellow pine or other hard wood that turns well on the lathe. The strips are 1.55 inches wide and .775 thick. This, unfortunately is .025-inch larger than a standard "1-inch" soft wood plank, but the extra thickness is required to properly fit the ¼-inch bolts. If you must use a soft wood plank, the extra thickness may be added to the pattern by adding a coat of Bondo auto body filler.

Locate dowels or "nail" type pins as described for the vibrator core box of *Volume 1*. About 1-inch from one end, clamp the strips together using a drywall screw. This end is later mounted in the lathe. A second screw may be located at the opposite end or at the large end of the body.

Mount and turn the pattern in the lathe. The large fillet at the front of the pattern is roughed out in the wood and covered with Bondo. Finish it using a coarse round file and sandpaper, while the pattern is turning.

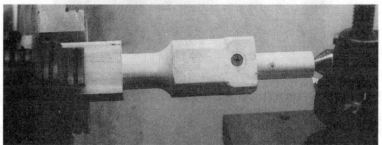

An inch or so away from the pattern, on one end, turn a 1-inch diameter boss with a ½-inch diameter by ¾ inch long dowel (see drawing). Move the pattern to a drill press or mill and using an end mill, cut the flats to mount the bosses. Using a ½-inch diameter end mill, drill the boss cutout at the parting line to accept the split boss pattern. Turn the pattern over and cut a 3/8-inch deep slot to accept the rib that attaches the top. Using a ¼-inch end mill, make the slots to accept the back and side ribs. All of the slots might also be cut using a table saw.

Mount a solid section of wood in the lathe and turn two additional solid bosses. Glue and clamp the bosses in place paying particular attention to the split boss on the parting line. Be sure the parting lines are aligned or later, you will not be able to separate the pattern.

To prevent glue from seeping between them, place a strip of wax paper between two ¼-inch thick strips of wood. With the pattern still clamped or screwed together, glue and insert the two ¼ inch thick strips to form the top rib. Cut, glue and insert the back ribs into their slots.

Cut and sand the draft on two strips for the top. Open the pattern, glue and clamp these strips in place. When the pattern is dry, fillet and sand the required draft. Note that the finished pattern is seen in the rubber mold chapter.

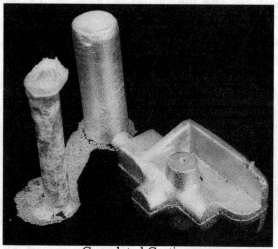
Completed Casting

Machining the Casting: Mount the casting in a 4-jaw chuck to that the rear of the casting is centered. Cut the rear flat and center drill. Using an 11/32-inch drill, drill completely through the casting. Use WD-40 or kerosene to lubricate the drill and clear the chips frequently. Enlarge the hole using a 23/64-inch bit.

Step drill the casting 2½ inches deep and 5/8-inch diameter using oil on the last pass to prevent aluminum from building up on the bit and making a rough and oversized cut.

Step-drill the casting to 31/32-inch diameter and 2-inches deep. Follow this with a well-oiled 63/64-inch bit. Insert a boring bar to the bottom of the 2-inch deep hole and make a cut .115 inches deep and .25-inch long (or wide, depending upon your orientation). Be careful not to gouge the end of the drilled hole because it forms the valve seat. The boring bar may be ground to form. Or you may make this cut in steps, about .025-inches deep per pass with a .015-inch deep finishing cut.

Using an oiled 3/8-inch reamer, ream the 23/64-inch hole. Finish the body using an oiled 1-inch diameter reamer. Turn the chuck by hand or lightly tap the momentary switch while feeding the reamer.

Move the casting to the drill press or mill and center drill the location of the air inlet port on the boss. It should be approximately 1.85 inches from the rear. Drill using a #5 drill. Drill and tap the pipe threads .625 inches deep. Center the drill in an outlet boss, drill and tap the air outlet, either left or right as needed.

Turn down one end of a short section of 1-inch diameter rod. Insert the small end into the drill chuck and slip the casting over the large end. This aligns the bore of the casting with the spindle. Clamp or hold in a vise. Loosen the drill chuck and remove the rod from the casting. Using a #4 bit, Drill then tap the holes for the ¼-inch bolts that hold the rear cover in place. These are located .54-inch off center in both the x and y axis.

Making the Piston: Make the piston from a 1.5-inch long section of 1-inch diameter rod. Chuck the rod in the lathe and drill ¾-inch deep using a #4 bit. Make facing cut to .9-inch diameter and .1-inch deep to hold the valve washer. Tap the hole ¼-20. Turn the piston around in the lathe and cut down a section for the spring holder. It should be .5-inches long and .5-inches in diameter. These values may be adjusted to suit different springs. I am using a **valve spring** from a Briggs & Stratton 3.5 horsepower engine. I have a good supply of these because I melt scrap engines. If you are melting scrap you will soon have a good supply too. You may use any spring as long as it fits the bore without binding and is stiff enough so that the valve actuates only when deliberate pressure is applied to the knee-pad.

Making the Valve Washer: Initially, I planned to use a 5/8 rubber faucet washer, however I could only find one in my scrap pile and could find only hard plastic washers at the local hardware store. The situation was remedied by making a washer cutting tool and cutting them from a section of automotive heater hose. Make the cutting tool from a section of 1-inch diameter rod. Drill and cut a hole that is .9-inches in diameter. Using a file, taper the outside edge of rod to a sharp point. Cut a strip of heater hose and lay it flat on a wood block. Set the cutting tool on top and give it a sharp whack with a hammer. It should cleanly cut a rubber disk. The ¼-inch diameter center hole is cut using a purchased hole punch, or you can make one as described above. The valve is secured to the piston using a ½-inch long ¼-inch bolt.

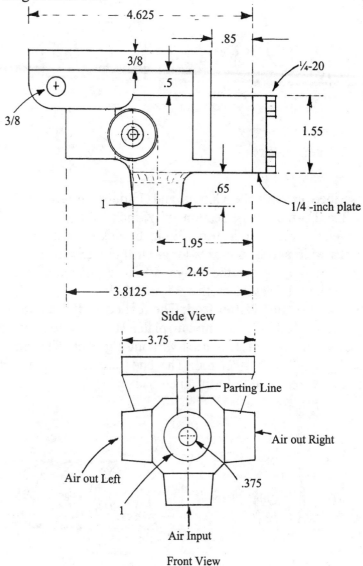

Side View

Front View

Making the Knee Pad: A 3 ½-inch diameter disk is torch cut from a section of 4-inch diameter pipe using the circle cutting attachment. The circle cutting attachment is

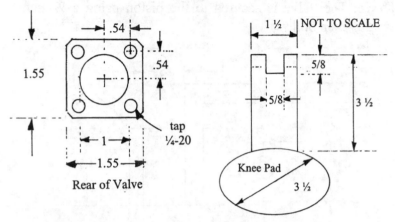

Rear of Valve

Knee Pad

described in both the Cupola and Tilting Furnace books. Drill a 1½-inch long section of ¼-inch pipe to 3/8-inch diameter. Weld it to a strip of ¼ inch thick steel strap 1½ - inches wide and 3 ½-inches long. Drill and saw or grind a notch in the top of the hinge 5/8-inch wide and deep. Secure it to the casting with a 3/8-inch bolt.

Rear Cover and Piston Rod: The rear cover is made from a section of ¼-inch aluminum plate. However it may be cast and turned flat. It is used without a gasket. The rod is cut from 3/8-inch steel rod. The ends are chamfered and dressed as required to fit the knee-pad.

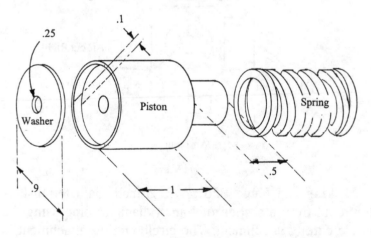

Washer   Piston   Spring

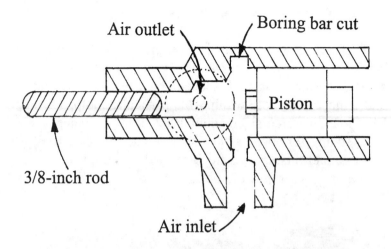

Air outlet   Boring bar cut

Piston

3/8-inch rod

Air inlet

Relative Locations of Ports

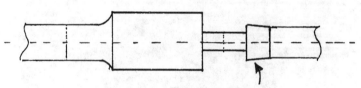

Turn the split boss when
turning the valve body

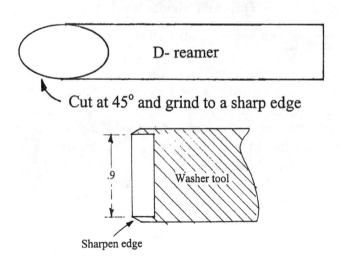

D- reamer

Cut at 45° and grind to a sharp edge

.9   Washer tool

Sharpen edge

MAKING A CYLINDER HEAD:

I often get odd repair jobs from the local technical school. On this occasion, an old cast iron cylinder head arrived that had a large L shaped crack that ran the length and almost the whole width. This particular motor has not been manufactured since about 1967. Used cylinder heads were selling for about $700. Because only a single head is required and this job is a freebie, I will use the lowest cost materials. The project did not appear to be too difficult for a single head.

Original Cylinder Head

There will be some solidification shrinkage, probably between .062 and .150 inch on the long side. If I locate all the finish machine work from the center bolthole, I can probably work from the original casting. Meaning, I will not make a complete new pattern.

The water jacket core is the most difficult part of the project so I'll start with that. Using a transfer punch on a section of ¼-inch Masonite, I mark the locations of all the bolt and sparkplug holes. I need to establish the location of the combustion chambers and their height. After coating the combustion chambers with petroleum jelly and a few strips of wax paper, dowels are inserted into the sparkplug holes and a plaster model of each combustion chamber is made.

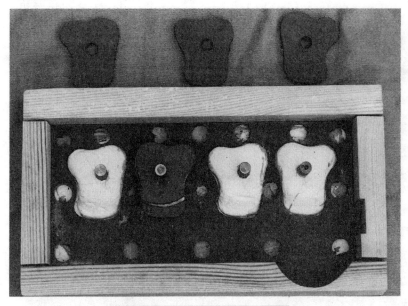

Roughing out the Core Box

Another wooden box is built and fitted to the outside of the head. Remember to subtract the wall thickness of both the top and bottom of the cylinder head from the depth of the box.

The sparkplug holes are transferred to the bottom of the box and the location of all the bolt holes are drilled in the Masonite. Using a head gasket, the shape of each combustion chamber is traced onto the Masonite and cut out using a scroll saw. The plaster models of the combustion chambers are transferred to the section of Masonite using the dowels for reference. In retrospect, it probably would have been fine to locate them using only the cutouts. The combustion chamber cut outs are sawn and sanded to conform to the shape of the plaster models and laid over top of them. Dowels are inserted into the bolt

161

hole locations and strips of Masonite are fitted between them creating the .25-inch wall thickness of the casting. The whole assembly is given several coats of Bondo and heavily filleted.

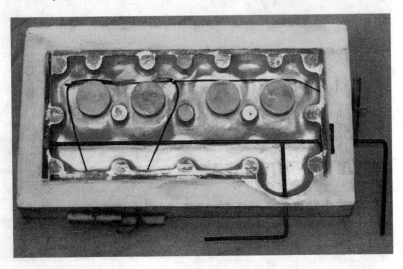

Completed core box with vent rods and string vents.

The casting walls appear to be between .25 and .375-inch thick. It is 13-inches long, 6.75-inches wide, 1.65 inches high and weighs 15 pounds. I will have to add 1/8-inch on the bottom surface and a few other surfaces for machining allowance. I chose 1/8-inch because that is a standard thickness for Masonite. The bolt bosses are also built up using 1/8-inch thick sections of a dowel.

Because the casting is fairly short, I will try to make another using the original casting. There will be some shrinkage, so I build up the pattern using a few coats of Bondo. This shrinkage and machining allowance adds about 3 pounds to the casting. (Iron weighs .26 pounds per cubic inch.)

13-inches x 6.75-inches x .125-inches x .26(lbs. / inch^3) = 2.85 lbs.

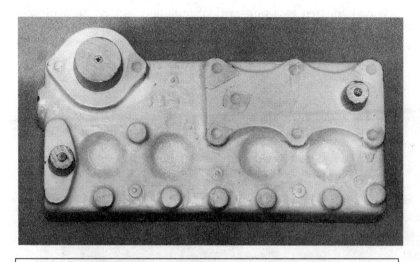

Built up head. Three core prints are added. Corresponding prints are also glued to the baked core.

A mold was made and the height of the core was checked. It was discovered that one print was a little off. Rather than make a new print, the core was shimmed by placing a penny under the short print. The core is secured using bent sections of welding rod inserted through a few bosses that will be machined out in the finished casting. A test casting is made and sawn in half to check the wall thickness. Every thing looks good except the core is a little gassy. Longer baking and a switch to wheat paste allows a reduction in the amount of binder, reducing the amount of gas. Note that all cores require a little cleaning with a file before use.

A few molds are made (both in iron and aluminum) and filter screens are used. The head is cast both in the cope and drag for evaluation. The cope casting has no visible inclusions at all. Machining reveals perfectly clean metal. The drag casting was easier to make, however a few inclusions were found. Machining removed all the surface flaws and revealed one very small inclusion in the casting.

It was located in a non-critical area and caused no problem in the finished casting.

Checking the core clearance in a test mold

I use two runners, one on each side of the casting so I will estimate that the total length of runners and gates is 36 inches. (The following figures are for iron) These probably weigh about 10 pounds. Although they may not be required, I will add four 2-inch diameter risers each one weighing approximately 4 pounds. The casting weight poured is 15 pounds for the head, 3 pounds of additional metal for machining, 16 pounds in risers and 10 pounds in the gating system. The complete estimated casting weight is 44 pounds. Later, it was found that the risers were not required.

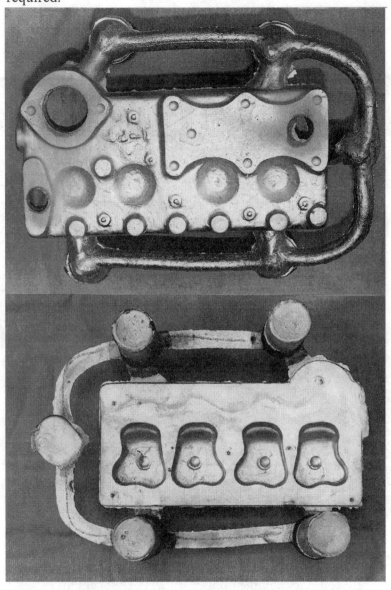

Completed Cylinder Head Casting

The top of the casting is surfaced and the sparkplug holes are faced using an end mill.

The sparkplug holes are drilled, reamed and tapped. A chamfer is cut at the top of each hole.

The casting is turned over, clamped to the mill and the bottom surface faced. The gasket is carefully centered between the combustion chambers The centermost bolt hole is drilled and used as a reference for all the other bolt holes in the head. The water-jacket cut outs are located by hand by pinning the gasket to the head and scribing the required cut out. These are finished using an end mill. The locations of all the remaining holes are carefully measured on the original head, relative to the center hole. By locating *everything* relative to the center of the casting, the shrinkage is averaged between the end points of the new casting. The locations are transferred to the head and all of the bolt holes are drilled.

Scribing the location of the water jacket cutouts.

To check the fit of the head, loose valves are coated with bearing blue and inserted into their guides. The head is fit onto the block and pinned in place using a few loose bolts. Each valve is pushed up and the head is removed. The head is inspected for any transferred bearing blue. One very small spot is detected and approximately .020-inch is ground off that location. The test was repeated and no bluing was transferred. The head was now complete and the cost of materials was negligible.

Completed Cylinder Head

CASTING PISTONS:

Modern pistons are cast in permanent molds, however many older engines used sand-cast pistons. Casting into permanent molds gives the aluminum a finer grain structure and higher mechanical properties than sand-cast pistons. You may accommodate this reduction in strength by heat treatment or by making the walls and pin bosses a little thicker. If the original wall thickness is .1-inch, you may increase it to .15 or more. The total weight may be a little higher but older engines are heavy and run at low speeds. It is unlikely that any imbalance will be detected.

In order to properly vent the core, cast the piston in the inverted position. Locate the core using a large print or disc around the base of the core. Form the core in a split core box and bake it in the upright position. Larger cores may slump and deform during baking and may have to be baked as split sections and glued together. The core should be

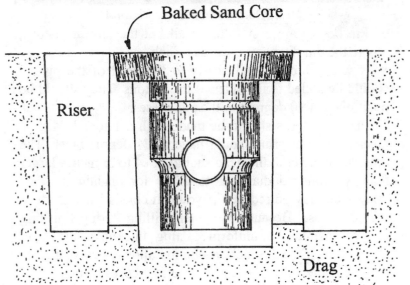

Molding a Piston in the Drag

well vented by pushing a vent rod through the green core or drilling through the center of a baked core. All cores should be checked for slump and squareness after baking. If you are using wheat flour for green strength and excessive slumping or gas is found, substitution of wheat paste (used for hanging wallpaper) should help eliminate the problem. Wheat paste is a much stronger binder, therefore less is required.

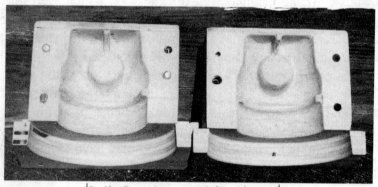
Split Core box used for Pistons

Pin bosses are cast solid on all but the largest pistons, where the pin diameter exceeds 1-inch.

Risers, which extend nearly to the head of the piston, should be added to the sides of the pistons. They should be positioned at 90 degrees to the pin bosses.

Piston patterns should be made with a 1 to 1 ½ degree taper. The core print should have a 7½ degree taper. The outside diameter at the head should be .4 to .5 inches larger than the finished diameter to allow for machining. This allowance may be reduced if you can consistently produce perfect cores. Because of center drilling during finishing and to maintain a uniform casting thickness, the head should have .2 to .25 inch machining allowance. Again this may be somewhat reduced with perfect cores. Single, free patterns may be rammed up, or they may be mounted to a match plate. Lower reject rates are found when using plate mounted patterns. Expect to pour a few test castings to check the patterns, risers and machining of the casting.

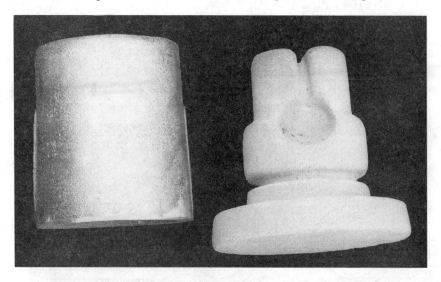

Completed Piston Casting (1930 Dodge) and Baked Sand Core

Left: Original Metric Piston from an Engine of Unknown Origin. Right: "Standard Size" Replacement Piston. Engine and connecting rod were bored to accept standard size.

X. AUTOMOTIVE CASTINGS

CASTING PISTON RINGS:

There are three primary ways to cast piston rings, however only one is still commonly used today.

Piston rings for steam engines are cast as a short length of pipe with a circular flange around one end for mounting the casting to a faceplate. Mounting tabs are unacceptable for this type of operation because stresses set up upon cooling of the casting causing the finished piston rings to warp.

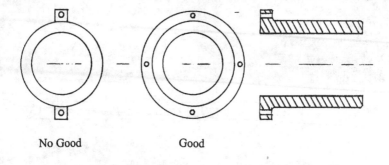

No Good   Good

Cast Stock for Turning Piston Rings as used on Steam Engines

Centrifugal casting is also used to produce stock for turning piston rings and in some cases, individual piston rings are cast. Molds may be horizontal as used in cast iron pipe or vertical as used in cylinder liners. Short horizontal molds may be inclined a few degrees (5°) so that the metal will flow to the back of the mold.

Rotating the mold forces the metal against the outer wall of the mold under increased pressure. This pressure may eliminate the need for a core as seen in iron pipe castings. It also causes impurities to separate and flow to the center of the mold as the metal solidifies. Because iron sulfides are lighter, higher concentrations will be found at the inner surface of the castings.

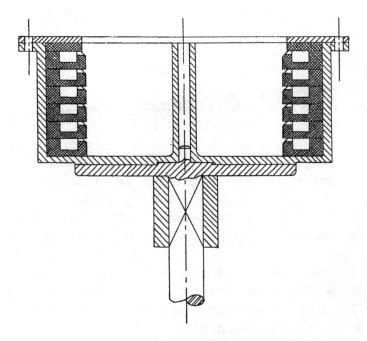

Vertical Centrifugal Casting with Baked Sand Cores

In order to prevent chilling of the metal, molds may be lined with sand and use sand cores. Iron molds use a mold coating, either refractory or silicon based. Coreless centrifugal molds require weighed charges for consistent wall thickness.

Typically, the centrifugal force used for this type of casting is 50 to 100 times the force of gravity (g). The required rpm is calculated as seen below with the optimal rpm being determined experimentally.

$$RPM = (265\sqrt{f\,d}) / d$$

f = force, number of "g"

d = diameter of mold in inches

173

Calculate the rpm needed for 50 g's in a 4-inch diameter pipe mold:

$$RPM = (265\sqrt{50 \times 4}) / 4 = 937 \, RPM$$

SAND CASTING PISTON RINGS:

Piston rings are commonly cast in thin molds stacked 18 to 20 high. Wooden patterns are made and sprung with a small wedge to the proper amount of gap. These patterns are poured in cast iron. The cast iron patterns are mounted individually on heavy cast iron plates in groups of 2, 4, 5, 6 or 8 depending upon the size and how far the iron will flow. The flask is closely formed around the patterns that are mounted around a central sprue. Flasks are typically 1.125-inches deep and have lifting handles mounted on either side. Each flask forms the cope of the one underneath it. The sprue in the bottom flask is plugged and forms the drag for the mold above it. In all the other molds, the sprue

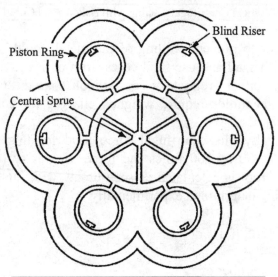

Close Fitting Flask used for Piston Ring Castings

is left open to form a continuous passage for the metal from the top to the bottom of the stack. The top flask has a pouring basin.

Where the two streams of molten metal meet opposite the gate, the iron is cool and holds more carbon in the combined form. Therefore a small blind riser opposite the gate is needed to prevent the metal from forming a hard spot. This riser serves as a heating element to prevent chilling of the far side of the ring. For automotive type rings, the average size is 1-inch long, ¼-inch wide and ½-inch high.

Details of Piston Ring Mold

Exceedingly hot metal is needed for such thin castings. The technique is similar to stove plate casting where molders line up on either side of the cupola spout and fill their ladles under a constant stream of molten iron. The ladles have hinged covers that are pulled back with a small chain and held open with a counterweight while they are under the cupola spout.

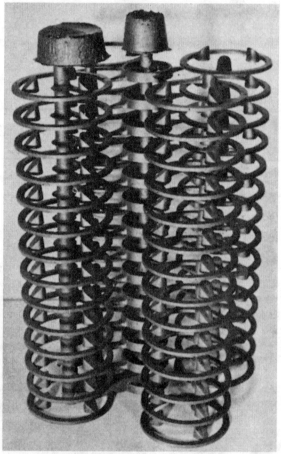

Completed Piston Ring Casting

During shakeout, the sprue is attached to a small hoist and vibrated. The flask sections fall off and the sand falls away. The rings are removed by running a gloved hand from top to bottom of each section of rings. The central sprue is returned to the cupola. Typical casting yield (ratio of metal poured to useable product) is 25%. Or, for every 100 pounds of metal poured, 25 pounds are useable product.

Iron is melted in 30 and 42-inch cupolas where the charges are 5.7 pounds of iron per pound of coke. Dolomite is added at the rate of .0625 pounds per pound of iron.

Sand is closely controlled as the slightest variation sends the castings to the scrap pile. Grains size is 150, moisture is 4 to 4.5%, and the clay content is 9%.

PRODUCTION OF THE FORD FLATHEAD V-8:

In 1934, the Ford V-8 plant produced 3500 to 4000 engines per 8-hour day and approximately 10,000 on a 24-hour shift. By using many extremely accurate gages, fixtures and careful control of both foundry practice and cupola melting, loss was held below 2 per cent.

Such a low loss rate requires that the cores and molds be accurately measured and deemed to be either perfect or not perfect. They are immediately discarded if not perfect. Production is planned to eliminate any gap between the furnace and the mold.

Molds for the camshafts are made from split patterns with the halves mounted on individual plates. Copes and drags are molded on separate machines. Each mold contains two castings connected by gates which deliver metal through shrink bobs at one end of the cope. A runner

feeds these gates from a central sprue. A smaller gate at the end of the casting is connected to a large riser located midway. Iron for camshafts comes directly from the cupola.

The crankshafts (left) are cast vertically in groups of four using a stacked mold built up of 16 individual dry sand cores. Each casting weighs 80 pounds giving a total of 320 pounds for the group of four. A heavy central sprue and individual risers on each casting brings the gross weight to 420 lbs.

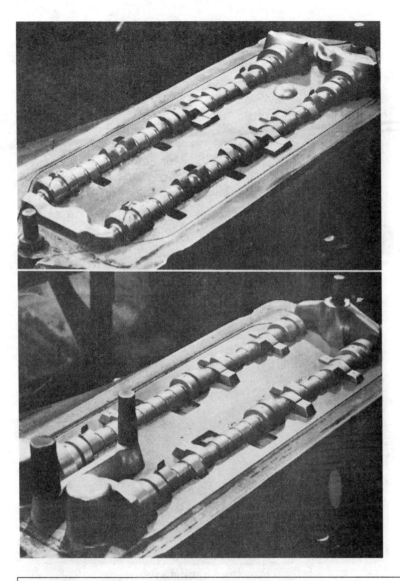

Camshaft Match-plates. Notice the core prints on the lobes having undercuts.

Stacked Crankshaft Molds

Iron for the cylinder blocks has a composition of: total carbon 3.20 to 3.40 per cent, silicon 1.90 to 2.10 per cent, sulfur 0.10 percent max, phosphorus 0.25 to 0.32 per cent, manganese 0.60 to 0.80 percent and copper 0.75 percent. The cupola charge is made up of pig iron and returns 85 percent and 15 per cent scrap steel.

Pouring Crankshafts

Cupola iron is transferred to an electric furnace where it is superheated to 2900 degrees F. The superheated metal is carried in 1500 pound ladles, sufficient to pour four cylinder blocks. Each mold requires approximately 300 pounds for castings gates and pouring basin. The finished cylinder block weighs 190 pounds. Risers at each end of the crankcase weigh 10 pounds. These risers, later found to be unnecessary, were eliminated saving approximately 40,000 pounds of metal per 8-hour shift.

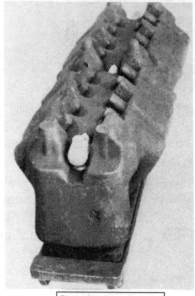

Assembling the Cylinder Cores

Cylinder cores are rammed from old sand and a pitch binder. A perforated steel arbor and a steel pipe are forced down through the center of the core. The core is blackened and dried in an oven for 2 hours and 20 minutes. The water jacket cores are made from a fine grade of lake sand bonded with oil. Pushing a steel wire through the cores makes the vents. A template covers the core vents and a thin coat of paste is applied to the adjoining faces. The cores are then bolted together through holes formed during molding. The bolt heads are covered with talc moistened with oil. The cores are then returned to the oven to dry. To prevent scratching the mold, strips of formed sheet metal cover the mold surfaces while the cores are set in place.

Crankcase Cores

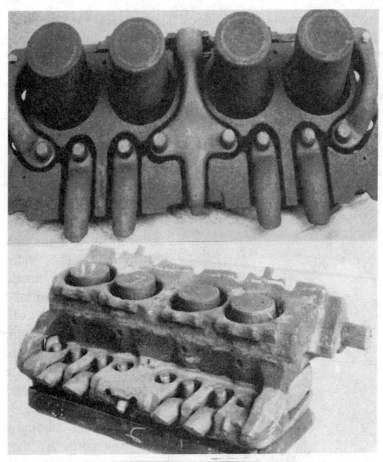

Cylinder Core Assembly

Core Checking Jig

Once the cores are set in place, the vents line up with openings in the side of the flask. The valve housing vents through the bottom and the

crankcase vents through the cope and one end of the flask.

Molds (left) for the cylinder blocks are made in cast steel flasks with all mating surfaces machined and ground. The pin holes are bored on a jig so that all flasks are interchangeable. The skin dried cope is guided into place using long pins which are removed and replaced with short pins which remain in place while the mold is poured. The short pins prevent the mold from shifting while it is being clamped or moved to the pouring station. To prevent the bottom plate from being pulled up too tight, a thin shim is set at each corner of the flask. Tolerances on the casting are held to 0.010 inch.

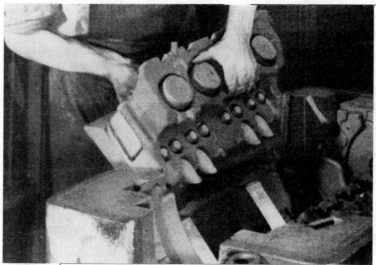

Setting the Completed Cores into the Mold

Iron is poured into a green sand pouring basin. A small sheet of paper covers the sprue opening in the cope before the pouring basin is set in position. This prevents any dust or dry sand from entering the mold with the first iron.

Completed V-8 Block Casting

Valve seats (below) are cast from high speed tool steel with a composition of: tungsten 14 to 17 percent, chromium 2.5 to 3.5 percent, carbon 1.20 to 1.40 percent, sulfur and phosphorous 0.05 percent, copper 1.50 to 2.00 percent, and silicon 0.30 to 0.60 percent. The seats are arranged 36 on a plate and are molded in shallow flasks. The cope mold is a flat plate. The molds are stacked 3 high for a total of 108 seats poured from a central sprue.

Valves are cast using split patterns mounted on flat plates in two opposing rows. They are gated from the ends with a long runner in the center. Electrically melted steel with a composition of: carbon 0.95- 1.20 per cent, silicon 2- 3.5 per cent, sulfur and phosphorous 0.05 per cent max, chromium 14.00- 16.00 per cent, and nickel 13 to 15 per cent.

XI. MISCLEANEOUS TOPICS:

MOLDING DEEP FINS:

   You are not likely to successfully mold thin deep fins in green sand. If the mold does not break when you are removing the pattern, the fins are likely to wash away when the mold is poured. The baked sand mold below uses approximately 750 pins to reinforce the sand between the fins. Using an aluminum pattern, you may bake the mold with the pattern in it, removing it when the mold has cooled. The aluminum pattern will expand during baking and shrink away from the mold wall when cool.

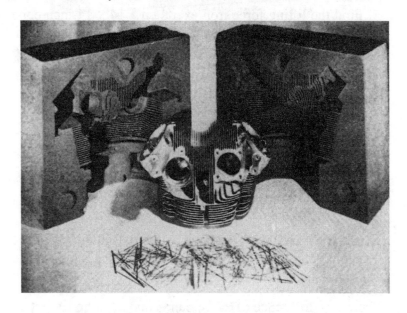

Approximately 750 pins reinforce the sand fins.

FINDING THE LIFTINGFORCE ON THE COPE:

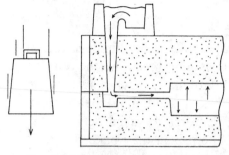

Flowing metal has energy like a falling weight. The momentum may lift a cope off of the mold, if it is not clamped or weighted. A mold is somewhat like a hydraulic cylinder. The liquid metal acts like hydraulic fluid and the upper casting surface acts like a piston to lift the cope as the metal is poured.

Find the lifting force iron exerts on a 14 x 14-inch cope if the casting is 8 x 10-inches and the head is 6-inches. Assume a 4-inch cope and that rammed sand weighs .06 pounds per cubic inch.

The approximate weight of the sand in the cope is:

$(14 \times 14 \times 4)$-inches^3 x .06 lb/inch^3 = 47 pounds

The lifting force acting on the cope is:

Area of casting x Head x Density of the Metal

$(8 \times 10 \times 6)$inches^3 x .26 lbs./inch^3 = 124.8 pounds

This is the pressure on the cope, without accounting for the dynamic pressure of the flowing metal, which would make this figure higher.

There is a difference of 77.8 pounds that must be weighted. A safety factor of 2 or 3 if the mold is poured quickly. The mold should clamped or have 150 to 200 pounds added to the cope. A general rule of thumb is to assume that the whole cope surface is under pressure and weigh it accordingly. This accounts for the safety factor.

$(14 \times 14 \times 6)$ x .26 = 306 pounds.

CONCLUSION:

There is much information contained in these two volumes. Fortunately, you do not have to know it all to get started.

As you can see from the cylinder head and piston projects, you may produce castings for virtually no cost, other than the time invested. With each new project you assume, you are learning. Learning can be a difficult and frustrating process; however if you work thoughtfully and persistently, success is almost inevitable. With each new success, your problem solving skills and creativity grow. Soon, you will surprise both your friends and yourself with what you can accomplish.

Pattern making is a valuable skill. Time spent making good patterns will repay you by making your molding much easier and your success rate much higher. Study both castings and molded plastic parts then ask yourself: "How did they make that? How would I make that? How did they get that pattern or plastic part out of the mold?" Always work towards the simplest solution. Anyone can make a part with a million dollars. It takes creativity and a little effort to make one for ten dollars.

If you are a beginner, don't be intimidated by the machine work. It is a skill that can be learned. Publications such as *Home Shop Machinist* will introduce you to what is available and spark your imagination. Machine-shop classes are offered by many community colleges and are a good value for the money. Regarding tooling, everyone starts with a Sears jig saw and a hammer. Over time, you can assimilate or build what you need.

As your foundry grows, you may build the larger furnaces in the series. You will soon find that melting larger amounts of metal is easy. However, you will find that keeping up with all of the sand required to pour all that metal is becoming a chore. Therefore, the foundry series continues with sand machinery and equipment.

Good luck and safe casting!

Steve Chastain

Bibliography

American Foundrymen's Society, Inc., *Aluminum Casting Technology*, 2nd ed., (Des Plaines, Illinois: American Foundrymen's Society, Inc., 1993).

American Foundrey Society Cast Metals Institute, *Basic Principles of Gating & Risering*, (Des Plaines, Illinois: American Foundrey Society Cast Metals Institute, 1973).

American Society for Metals, *Aluminum, Volume 1. Properties, Physical Metallurgy and Phase Diagrams*, (Metals Park, Ohio: American Society for Metals, 1967).

American Society for Metals, *Metals Handbook Ninth Edition, Volume 2: Properties and Selection: Nonferrous Alloys and Pure Metals*, (Metals Park, Ohio: American Society for Metals, 1979).

Backerud, Lennart, Guocai Chai and Jarmo Tamminen, *Solidification Characteristics of Aluminum Alloys, Volume 2, Foundry Alloys*, (Stockholm, Sweden: American Foundry Society,).

Campbell, OBE, FEng, John, *Castings*, (Oxford, Auckland, Boston, Johannesburg, Melbourne, New Delhi: Butterworth Heinemann, 1991).

Dwyer, Pat, *Gates and Risers for Castings*, 3rd ed., (Cleveland, Ohio: The Penton Publishing Company, 1949).

Gaskell, David R., *Introduction to Metallurgical Thermodynamics*, (Scripta Publishing Company, 1973).

Guy, Albert G., *Elements of Physical Metallurgy*, 2nd ed., (Massachusetts: Addison-Wesley Publishing Co., 1959).

Heine, Richard W., Carl R. Loper, Jr., and Philip C. Rosenthal, *Principles of Metal Casting*, (McGraw-Hill Book Company, Inc., 1967).

Kaiser Aluminum & Chemical Sales, Inc. *Casting Kaiser Aluminum*, 3rd ed., (Oakland, California: Kaiser Aluminum & Chemical Sales, Inc., 1974).

Lord, James Osborn, *Alloy Systems, An Introductory Text*, (New York and London: Pitman Publishing Corporation, 1949).

Malleable Founders Society, *Malleable Iron Castings*, (Cleveland, Ohio: Malleable Founders Society, 1960).

Porter, D.A. and K.E. Easterling, *Phase Transformations in Metals and Alloys*, 2nd ed., (1992).

Ruddle, M.A., F.I.M., R. W., *The Solidification of Castings*, (Belgrave Square, London: The Institute of Metals, 1957).

Shrager, A.B., B.S., Arthur M., *Elementary Metallurgy and Metallography*, 3rd ed., (New York: Dover Publications, Inc. 1969).

Smith, William Fortune, *Structure and Properties of Engineering Alloys*, (McGraw-Hill, Inc., 1981).

SUPPLIERS:

American Foundrymen's Society
505 State Street
Des Plaines, IL 60016-8399

Foundry books, from basic to advanced
800 537 4237 www.afsinc.com
moderncastaing.com

Hickman Williams
US 800 862-1890
Canada 800 265 6415
Mexico 011 52 8 363-4041

Freeman Supply
1101 Moore Rd..
Avon, Ohio 44011
800 321-8511
Pattern making supplies

Joy-Mark Inc.
2121 E Norse Ave.
Cudahy, Wis. 53110
414 769-8155
Ceramic reinforced ladles.
Part# 407   10 x 8.5 x 11.375

This ladle is very light. I use it for all my
iron and aluminum work. It is highly
recommended.

Insulating riser sleeves, riser toppings

Lindsay Publications
P.O. Box 538
Bradley, IL 60915

815 935 - 5353
lindsaybks.com
(Foundry books , Technical books)

Home Shop Machinist Magazine
2779 Aero Park Drive
Traverse City, MI 49686

Magazine for beginning and intermediate
machinists

Smooth-On
Easton, PA
www.smooth-on.com
(888) 765-9835
Casting Rubbers

Filters – Refractory Cloth:
Amatek
900 GreenBank Rd.
Wilmington, DE 19808
800 - 441-7777
www.ametek.com/haveg

Budget Casting Supply
60 East 40th Ave.
San Mateo CA 94403
650 345-3891
budgetcastingsupply.com

Piedmont Foundry Supplies
3191 Rogers Lane
Cloverdale, VA 24077
Phone 992-3911
Columbus,GA Phone 324-5938

Porter Warner Foundry Supplies: 205 251-8223
Metal Ingot – all types

Belmont Metals
belmontmetals.com
718 342 4900

Atlas Metal Sales
1401 Umatilla St.
Denver, CO 80204 - 2432
303 623 0143

Ferro silicon, Nickel, Copper Phosphorus:
ASi International 1440 E. 39th St., Cleveland, Oh 44114 (216) 391-9900

Harbor Freight Tools
1 800 423- 2567

Enco Tools
800 873-3626
www.use-enco.con

Aluminum Casting Alloys:

Alloy	Si	Fe	Cu	Mn	Mg	Cr	Ni	Zn	Ti	Sn
208	3	1.2	4	0.5	0.1	0	0.35	1	0.25	0
242	0.7	1	4	0.35	1.5	0.25	2	0.35	0.25	0
295	1.1	1	4.5	0.35	0.03	0	0	0.35	0.25	0
319	6	1	3.5	0.5	0.1	0	0.35	1	0.25	0
332	9.5	1.2	3	0.5	1	0	0.5	1	0.25	0
336	12	1.2	1	0.35	1	0	2.5	0.35	0.25	0
356	7	0.6	0.25	0.35	0.35	0	0	0.35	0.25	0
360	9.5	2	0.6	0.35	0.5	0	0.5	0.5	0	0.15
383	10.5	1.3	2.5	0.5	0.1	0	0.3	3	0	0.15
390	17	0.5	4.5	0.1	0.55	0	0	0.1	0.2	0
443	5	0.8	0.6	0.5	0.05	0.25	0	0.5	0.25	0
535	0.15	0.15	0.05	0.175	6.85	0	0	0	0.175	0
771	0.15	0.15	0.1	0.1	0.9	0.2	0	7	0.15	0
851	2.5	0.7	1	0.1	0.1	0	0.5	0	0.2	6

INDEX:

Air Valve, making, 152
Aluma-Kote, 40
Aluminum,
    Alloys 21
        Briggs & Stratton, 27
    Brittleness in aluminum, 23,33-34
    Bronze, 54
    Dross, 35
    Flux, 40
    Grain Refiners, 23
    Heat Treatment of, 25-26
    Hydrogen Solubility, 38
    Numbering System, 27
    Oxides, 32-34
    Phosphorous, addition of, 24
    Precipitation hardening, 25,26
    Rust, reaction with, 39
    Silicon, addition of, 22, 23
    Solidification Range, 30
Basin, Pouring, 77,79,80
Brass, 42, 44
Carbide, 65
Carbon, 68
    Equivalent, 70
Chill, 13, 56-58, 107
Chlorine, 40
Choke, 81
    Area, 88
Chromium, 72
Cincinnati Mill, 146, 147
Columnar, Grains, 8
Coping Out, 123
Copper,
    Deoxidizing, 63
    High Conductivity, 42-43
    Melting, 62
Core
    Prints, 124
    Sand, Copper alloys, 60
Cracking,
    Iron, 72
    Wheel, 126
Crankshaft, 127
Cylinder
    Fins, Casting, 184
    Head, Making, 160
Dendrite, 10
Dowel, 123
Draft, pattern 121
Ductile Iron, 74
Equi-axed, 8
Eutectic, 7

Freezing, 11
Extruder, Wax, 120
Feed Metal, restriction of, 8,11
Flaring, Zinc, 42, 63
Flux
    Aluminum, 40
    Iron, 75
Fillet, 119
Filters, 87
Fins, Cylinder, Casting 184
Flask Latch, 148
Follow Board, 122
Gate, Gating Systems, 77
    Bronze, 60
    Design of, 88
    Requirements of, 79
Graphite,
    Type, 65-66
Harley Davidson, 73
Heat of Fusion, 7
Horn Gate, 49
Ingate, 84-86
Iron,
    Class, 69
    Composition of, 67
    Cracking, causes, 72
    Hyper -Hypoeutectic, 13
    Types, 65
Ladle Wash 39,40
Latex, 129-130
Lead Sweat, 53
Lifting Force, On Cope, 185
Liquidus, 9
Long Freezing Range, 12,14,15
Manganese, 73
    Bronze, 47-50
    Sulfide, 72, 73
        Cracking caused by, 72
Micro porosity, shrinkage, 11
Mushy
    Alloy, 99
    Phase, 9
Nail Grate, 49
Nickel, 72
Nuclei, 8
Padding, 100
Pasty, solidification, 9
Pattern,
    Making, 117
    Split, 121, 123
Phase,7
    Diagrams, 16-20
Phosphorous,
    Aluminum, refinement by, 24

191

Iron, addition to, 71
Pipe, riser, 13
Piston,
   Making, 169
   Rings, 172
Plastic, Liquid, 138
Pouring Time, 88-92
Pistol, Casting Model of, 128
Precipitate, 7
Riser, 99
   Blind, 103
   Dimensions, 101-102
   Exothermic, 114
   Feeding Distance, 104, 105
   Heat Loss, 108-110
   Sleeves, 109, 111
   Topping, 113
   Types, 78
   Weight of, 106
Pouring,
   Cup, 77
   Height, Effect of, 33, 185
Rubber
   Molds, 129-130
   Silicone, 131
Runner, 82
Shrink Ruler, 117, 118
Shrinkage,
   Allowance, 117,118
   dispersed, 13

Silicon,
   Bronze, 45
      Wormy defect, 45
   In Iron, 69
Slag-Off, 76
Soda Ash, 75-76
Solid Solution, freezing, 10
Solidification Range, Copper, 58
Solidus, 9
Sodium
   Modification of aluminum, 12
Spokes, Wheel, Casting, 126
Sphere, Solid Casting, 55
Sprue, 80, 81, 92
Statuary Bronze, 51
Stop Off, 127
Sulfur,
   In iron, 71
Thickness, Section, Combining, 125
Tin Bronze, 50-52
Titanium Boron, 23
Trap, Dross, 82
V-8, Flathead, Casting, 178, 183
Vacuum Degassing, 134, 137-138
Wall Thickness,
   Iron castings, 69
Wax, Fillet Extruder, 120
Well, 92
White Iron, 65
Zinc Flaring, 42, 63

Additional titles available at: **StephenChastain.Com** where you will find color photos and a list of foundry, auto restoration and alternative energy titles

Email: SteveChastain@hotmail.com